NUCLEAR AND ATOMIC PHYSICS AT ONE GIGAFLOP

NUCLEAR SCIENCE RESEARCH CONFERENCE SERIES

A series of monographs edited by Alexander Zucker, Oak Ridge National Laboratory

VOLUME 1	**GIANT MULTIPOLE RESONANCES** Edited by Fred E. Bertrand
VOLUME 2	**CONTINUUM SPECTRA OF HEAVY ION REACTIONS** Edited by T. Tamura, J. B. Natowitz and D. H. Youngblood
VOLUME 3	**LASERS IN NUCELAR PHYSICS** Edited by C. E. Bemis, Jr. and H. K. Carter
VOLUME 4	**HIGH ANGULAR MOMENTUM PROPERTIES OF NUCLEI** Edited by Noah R. Johnson
VOLUME 5	**HEAVY IONS AND NUCLEAR STRUCTURE** Edited by B. Sikora and Z. Wilhelmi
VOLUME 6	**NUCLEAR PHYSICS WITH HEAVY IONS** **Prospects at Energies Below 20 MeV/amu** Edited by Peter Braun-Munzinger
VOLUME 7	**INSTRUMENTATION FOR HEAVY ION NUCLEAR RESEARCH** Edited by Dan Shapira
VOLUME 8	**COLLECTIVE MOTION AND GIANT RESONANCES** Edited by Z. Wilhelmi and M. Kicińska-Habior
VOLUME 9	**HEAVY IONS IN NUCLEAR PHYSICS** Edited by Z. Wilhelmi and M. Kicińska-Habior
VOLUME 10	**NUCLEAR PHYSICS: RECENT TRENDS AND DEVELOPMENTS** Edited by Z. Wilhelmi, G. Szelfińska, and M. Kicińska-Habior
VOLUME 11	**NUCLEAR FISSION AND HEAVY-ION-INDUCED REACTIONS** Edited by Wolf-Udo Schröder
VOLUME 12	**COHERENT EFFECTS IN HIGHLY EXCITED NUCLEI** Edited by Z. Wilhelmi and G. Szeflińska
VOLUME 13	**SYMMETRIES AND NUCLEAR STRUCTURE** Edited by R. A. Meyer and V. Paar
VOLUME 14	**PHYSICS AT LEAR WITH LOW ENERGY ANTIPROTONS** Edited by C. Amsler, G. Backenstoss, R. Klapisch, C. Leluc, D. Simon and L. Tauscher
VOLUME 15	**THEORETICAL AND EXPERIMENTAL METHODS OF HEAVY-ION PHYSICS** Edited by Z. Wilhelmi and G. Szeflińska
VOLUME 16	**NUCLEAR AND ATOMIC PHYSICS AT ONE GIGAFLOP** Edited by C. Bottcher, M. R. Strayer and J. B. McGrory

NUCLEAR AND ATOMIC PHYSICS AT ONE GIGAFLOP

Proceedings of the Nuclear and Atomic
Physics Conference at One Gigaflop
held at Oak Ridge, Tennessee
April 13–16, 1988

Edited by
C. Bottcher, M.R. Strayer, J.B. McGrory
Physics Division
Oak Ridge National Laboratory
Tennessee, USA

harwood academic publishers
chur • london • paris • new york • melbourne

Harwood Academic Publishers

Post Office Box 786
Cooper Station
New York, New York 10276
United States of America

Post Office Box 197
London WC2E 9PX
England

58, rue Lhomond
75005 Paris
France

Private Bag 8
Camberwell, Victoria 3124
Australia

Library of Congress Cataloging-in-Publication Data

Nuclear and atomic physics at one gigaflop / edited by C. Bottcher, M.R. Strayer, and J.B. McGrory.
p. cm.—(Nuclear science research conference series, ISSN 0250-4375; v. 16)
Papers from a workshop held at Oak Ridge, Tenn., Apr. 14–16, 1988.
Bibliography: p.
Includes index.
ISBN 3-7186-0492-2
1. Nuclear physics—Data processing—Congresses.
2. Supercomputers—Congresses. I. Bottcher, C.
II. Strayer, M. R. III. McGrory, J. B. IV. Series.
QC783.3.N83 1988
539.7′028′5—dc19 88-31829
CIP

Printed in the United States of America.

CONTENTS

Preface to the Series vii

Foreword ix

List of Participants xi

QED of Strong Fields and Delayed Heavy Ion Collisions
S. Graf, O. Graf, T. de Reus, U. Müller, J. Reinhardt,
B. Müller and W. Greiner 1

Precision Calculations of QED Effects in Atoms
J. Sapirstein 25

Spectroscopy and Strangeness in the Quark-Gluon Plasma
John B. Kogut 43

Green's Function Monte Carlo with "Shadow" Trial Functions
S. A. Vitiello, K. J. Runge and M. H. Kalos 57

Drilling Holes Through Nuclei, Helium Drops, and Fermi Gases
Steven C. Pieper 69

Numerical Solution of High Temperature Superconductor
Spin Systems
Ted Barnes 83

Many-Body Effects in Ion-Atom Collisions
R. E. Olson 107

Atomic Structure Calculations on the Cray X-MP
Charlotte Froese Fischer 117

Hadronic Structure
J. W. Negele 133

The Relativistic Mean-Field Model in Nuclear Physics
P.-G. Reinhard, V. Blum, J. Fink, M. Rufa, J. Maruhn,
W. Greiner, C. Bottcher, M. R. Strayer, S. Umar,
J. Wu and R. Y. Cusson 157

Hydrodynamic Model of High-Energy Heavy-Ion Collisions
Joachim A. Maruhn 173

Soliton Bag Models
Lawrence Wilets 193

Condensates in the Standard Model
Carl Shakin 207

A Tutorial on Distributed Memory Programs
D. E. Stevenson 221

High Performance Parallel Computers for Science:
New Developments at the Fermilab Advanced Computer Program
T. Nash, H. Areti, R. Atac, J. Biel, A. Cook, J. Deppe,
M. Edel, M. Fischler, I. Gaines, R. Hance, D. Husby,
M. Isely, E. Miranda, M. Miranda, T. Pham and T. Zmuda 229

Computers and Molecular Quantum Mechanics: 1965–1988,
a Personal Perspective
Henry F. Schaefer, III 247

Ab Initio Calculations of Processes in Small Molecules
Kate P. Kirby 269

Nonperturbative Approaches to Atomic and Molecular
Multi-Photon Processes in Intense Laser Fields
Shih-I Chu 283

The Nuclear Few-Body System
G. L. Payne 305

Author Index 325

For multi-author papers, the speaker is underlined.

PREFACE TO THE SERIES

Nuclear science is now in full flower. From obscure origins in a few university laboratories it has grown to be one of the richest fields of scientific endeavor in the history of science. It has spun off two virtually unlimited sources of energy for mankind, fission and fusion. It is used broadly in many applied areas that range from medicine to agriculture to mining to metallurgy, and withal it continues to be a happy and productive basic science. New approaches, new particles, new energy regimes, and above all new ideas continue to provide astonishing theoretical insights and frequently unexpected experimental results.

Burgeoning growth is a healthy sign, but it carries with it a concomitant information explosion: as the science expands, the number of scientists increases, and specialized areas appear where once there existed but one unified field. To be sure, the technical literature performs its task adequately, but if a practitioner in a narrow field of nuclear science is to be aware of everything that he needs to know, he must read a dozen or more journals. Frequently a year or more elapses before a physicist's results appear in the journal literature. Because of such formidable information barriers, rapid communication in science is now principally oral rather than written. With this realization, specialized scientific conferences become a prime arena of scientific exchange and scientific publication. Spectacular results are now nearly always first communicated at conferences where they also receive their first baptism under fire. Sometimes new ideas may die at the same conference where they were first revealed. Conferences provide the stage where experiment and theory are joined, where results of different groups are compared, and new syntheses are evolved.

A conference will frequently broaden its program to include a paper or two on ancillary but essential developments. For example, a meeting on heavy-ion nuclear reactions might well include a session on new accelerator ideas or ion source developments. Finally, a conference ends with a summary by a noted practitioner in the field, and this summary can have great and lasting value for the reader.

It is for all these reasons that we have undertaken to publish a conference series in nuclear science, NUCLEAR SCIENCE RESEARCH. We expect this series, while it specializes in the areas of nuclear physics, to cover topics both basic and applied. We expect each conference to have its specialized nature so well defined that the proceedings will serve as a reference volume of lasting importance, and we expect that the resource provided to the working physicist by this series will enable those who were not participants to get both the data and the flavor of the conference. Speed of publication is essential for such proceedings. We are told that with modern printing methods the book can appear 3 to 4 months after receipt of the last manuscript. It is our intent that the series NUCLEAR SCIENCE RESEARCH publish proceedings of important conferences in the field, and our hope that it will provide a record of advances in a way that is most useful to the working scientist.

Alexander Zucker

FOREWORD

A three-day workshop on problems in atomic and nuclear physics which depend on and are, at present, severely limited by access to supercomputing at effective rates of one gigaflop or more, was held at Oak Ridge, Tennessee, April 14–16, 1988.

The participants comprised 70 researchers from universities, industries and laboratories in the United States and Europe. In this volume are presented 19 invited talks from that meeting on atomic and nuclear physics topics and on modern parallel processing concepts and hardware. The physics topics included strong fields in atomic and nuclear physics, the role of quarks in nuclear physics, the nuclear few-body problem, relativistic descriptions of heavy-ion collisions, nuclear hydrodynamics, Monte Carlo techniques for many-body problems, precision calculation of atomic QED effects, classical simulation of atomic processes, atomic structure, atomic many-body perturbation theory, quantal studies of small and large molecular systems, and multi-photon atomic and molecular problems.

Other talks dealt with present or proposed multi-processor computers. In virtually every field it was stressed that progress will only be possible with ready access to computing, not only in the gigaflop range (now available to a privileged few), but in the teraflop range. More than half the talks discussed variants of Monte Carlo techniques. The general consensus was that such stochastic calculations are ''embarrassingly'' easy to formulate as parallel algorithms. An important future research area will be the construction of parallel algorithms for this and other problems.

The techniques of computational quantum chemistry have many industrial applications ranging from surface chemistry to pharmacology. It was estimated that in the not-too-distant future 70% of supercomputer time in the U.S. will be consumed by industrial chemical applications. From the hardware perspective, the fabrication of custom-designed chips is becoming so reliable and economical that in many fields where calculations are dominated by a particular technique, it will be far more profitable to design and build specialized parallel processors than to write programs which run on a general-purpose computer.

The conference revealed an extensive overlap between the interests of atomic and nuclear physicists in computational techniques, and illustrated once again the benefits derived from cross-fertilization of different branches of physics. While the topics discussed were essentially basic research problems, many speakers referred tangentially to applications in applied areas such as fluid mechanics and information processing. In informal discussions much interest was expressed in the implications of the supercomputing revolution for the United States' industrial competitiveness, and in the role basic researchers in universities and laboratories can play in training people in this area. As regards facilities and resources, no single solution is possible in terms of architecture or organization (centralized vs. local facilities). What is urgently required is improved national networks so that everyone can be matched to the right facilities regardless of geographical accidents.

We would like to acknowledge with gratitude the hours of work devoted to the conference by the conference secretary, Ms. Althea Tate. We are grateful to Dr. Paul Huray, Director of the University of Tennessee Science Alliance, Dr. William Felling, Executive Director of the Oak Ridge Associated Universities, and to Dr. James B. Ball of the Physics Division of the Oak Ridge National Laboratory for the generous support these organizations provided in sponsoring this meeting. Finally, we express our appreciation to Digital Equipment Corporation and Floating Point Sytems for their corporate sponsorship of social events at the meeting.

The Organizers,

C. Bottcher
M. R. Strayer
J. B. McGrory

PARTICIPANTS

Adams, Jeannette, University of Maryland, College Park, Maryland, U.S.A.
Alam, Mohammed M., Southern Illinois University, Carbondale, Illinois, U.S.A.
Baker, Howard C., Berea College, Berea, Kentucky, U.S.A.
Barnes, Ted, University of Toronto, Toronto, Ontario, Canada
Becker, Richard L., Oak Ridge National Laboratory, Oak Ridge, Tennessee, U.S.A.
Bottcher, C., Oak Ridge National Laboratory, Oak Ridge, Tennessee, U.S.A.
Bottrell, Gerald, Oak Ridge National Laboratory, Oak Ridge, Tennessee, U.S.A.
Chu, Shih-I, University of Kansas, Lawrence, Kansas, U.S.A.
Cusson, R.Y., Science Applications International Corp., Los Altos, California, U.S.A.
Darewych, J.W., York University, Toronto, Ontario, Canada
Davies, K.T.R., Oak Ridge National Laboratory, Oak Ridge, Tennessee, U.S.A.
Dean, David, Oak Ridge National Laboratory, Oak Ridge, Tennessee, U.S.A.
deForcrand, Philippe L., CRAY Research, Inc., Mendota Heights, Minnesota, U.S.A.
Dubé, Louis J., University of Laval, Quebec, Quebec, Canada
Dworzecka, Maria, George Mason University, Fairfax, Virginia, U.S.A.
Ernst, David, Texas A&M University, College Station, Texas, U.S.A.
Froese Fischer, C., Vanderbilt University, Nashville, Tennessee, U.S.A.
Guidry, M. W., University of Tennessee, Knoxville, Tennessee, U.S.A.
Hammarén, Esko, Centre for Scientific Computing, Espoo, Finland
Harrison, Don E., Naval Postgraduate School, Monterey, California, U.S.A.
Heath, M. T., Oak Ridge National Laboratory, Oak Ridge, Tennessee, U.S.A.
Hicks, H. Richard, Oak Ridge National Laboratory, Oak Ridge, Tennessee, U.S.A.
Johnson, Elijah, Oak Ridge National Laboratory, Oak Ridge, Tennessee, U.S.A.

Kahane, Sylvian, Oak Ridge National Laboratory, Oak Ridge, Tennessee, U.S.A.

Kalos, M. H., New York University, New York, New York, U.S.A.

Kelly, H. P., University of Virginia, Charlottesville, Virginia, U.S.A.

Kirby, Kate, Harvard-Smithsonian Center for Astrophysics, Cambridge, Massachusetts, U.S.A.

Kogut, John B., University of Illinois, Urbana, Illinois, U.S.A.

Kumano, Shunzo, University of Tennessee, Knoxville, Tennessee, U.S.A.

LaGattuta, Kenneth, Los Alamos National Laboratory, Los Alamos, New Mexico, U.S.A.

Liang, Yigao, University of Kentucky, Lexington, Kentucky, U.S.A.

Maienschein, Fred, Oak Ridge National Laboratory, Oak Ridge, Tennessee, U.S.A.

Maruhn, Joachim A., University of Frankfurt, Frankfurt, F.R.G.

McGrory, J. B., Oak Ridge National Laboratory, Oak Ridge, Tennessee, U.S.A.

Müller, Berndt, University of Frankfurt, Frankfurt, F.R.G.

Munro, John K., Oak Ridge National Laboratory, Oak Ridge, Tennessee, U.S.A.

Nash, E. Thomas, Fermi National Accelerator Laboratory, Batavia, Illinois, U.S.A.

Negele, J. W., Massachusetts Institute of Technology, Cambridge, Massachusetts, U.S.A.

Nestor, C. W., Oak Ridge National Laboratory, Oak Ridge, Tennessee, U.S.A.

Obenshain, F. E., Oak Ridge National Laboratory, Oak Ridge, Tennessee, U.S.A.

Oberacker, Volker E., Vanderbilt University, Nashville, Tennessee, U.S.A.

Olson, R. E., University of Missouri, Rolla, Missouri, U.S.A.

Payne, Gerald L., University of Iowa, Iowa City, Iowa, U.S.A.

Pieper, Steven C., Argonne National Laboratory, Argonne, Illinois, U.S.A.

Pindzola, Michael S., Auburn University, Auburn, Alabama, U.S.A.

Pinkston, W. T., Vanderbilt University, Nashville, Tennessee, U.S.A.

Plasil, F., Oak Ridge National Laboratory, Oak Ridge, Tennessee, U.S.A.

Ray, John R., Clemson University, Clemson, South Carolina, U.S.A.

Reinhard, P.-G., University of Erlangen, Erlangen, F.R.G.
Sapirstein, Jonathan, University of Notre Dame, Notre Dame, Indiana, U.S.A.
Satchler, G. R., Oak Ridge National Laboratory, Oak Ridge, Tennessee, U.S.A.
Schaeffer, H. F., University of Georgia, Athens, Georgia, U.S.A.
Schlifer, Ian, York University, Toronto, Ontario, Canada
Shakin, C. M., Brooklyn College, Brooklyn, New York, U.S.A.
Sorensen, Soren P., University of Tennessee, Knoxville, Tennessee, U.S.A.
Stauffer, A. D., York University, Toronto, Ontario, Canada
Stevenson, D. E., Clemson University, Clemson, South Carolina, U.S.A.
Strayer, M. R., Oak Ridge National Laboratory, Oak Ridge, Tennessee, U.S.A.
Symmes, W. K., Floating Point Systems, Inc., Asheville, North Carolina, U.S.A.
Umar, A. S., Vanderbilt University, Nashville, Tennessee, U.S.A.
Vallieres, Michel, Drexel University, Philadelphia, Pennsylvania, U.S.A.
Ward, Richard C., Oak Ridge National Laboratory, Oak Ridge, Tennessee, U.S.A.
Ward, Robert C., Oak Ridge National Laboratory, Oak Ridge, Tennessee, U.S.A.
Ward, Sandra, York University, Toronto, Ontario, Canada
Weinstein, John, University of Tennessee, Knoxville, Tennessee, U.S.A.
Wilets, Lawrence, University of Washington, Seattle, Washington, U.S.A.
Wooten, John W., Oak Ridge National Laboratory, Oak Ridge, Tennessee, U.S.A.
Wu, Jianshi, Oak Ridge National Laboratory, Oak Ridge, Tennessee, U.S.A.
Young, Glenn, Oak Ridge National Laboratory, Oak Ridge, Tennessee, U.S.A.
Wong, Cheuk-Yin, Oak Ridge National Laboratory, Oak Ridge, Tennessee, U.S.A.

QED OF STRONG FIELDS AND DELAYED HEAVY ION COLLISIONS

S. GRAF, O. GRAF, T. DE REUS, U. MÜLLER, J. REINHARDT, B. MÜLLER, W. GREINER
Institut für Theoretische Physik, Joh. Wolfg. Goethe-Universität, Postfach 111932, D-6000 Frankfurt am Main 11, West Germany
G. SOFF
Gesellschaft für Schwerionenforschung, Postfach 110552, D-6100 Darmstadt, West Germany

Abstract We review methods of measuring nuclear reaction times through processes of quantum electrodynamics in strong fields. Possible methods for the detection of spontaneous positron production are also discussed.

INTRODUCTION

Quantum electrodynamics (QED) of strong fields[1] in collisions of very heavy ions has continued to fascinate theoreticians and experimentalists alike for more than a decade. Most notably, the predicted process of spontaneous positron creation in strong electric fields has been a strong source of motivation to carry on research in this field, as have been the still puzzling positron and electron-positron coincidence peaks observed at GSI.[2] Although much work has been done during the past three years, these phenomena are not yet understood.

A second aspect of QED of strong fields which has been less well publicized, but which has already borne fruit, is that electrons and positrons emerging from the heavy ion collision carry information on nuclear reaction times.[3] As a matter of fact, the leptons may be considered as a kind of "atomic clock" for the nuclear reaction that ticks at intervals of about 10^{-21} s, because the electronic binding energies involved here are of order of 1MeV. Experimentally measured electron spectra have provided the first model-independent determinations of interaction times in deep-inelastic nuclear reactions,[4] and they promise to be an equally useful tool in the energy region around 100MeV per nucleon.

In this report we will first briefly review the computational formalism for QED of strong fields, in particular, including a nuclear time delay. We will then compare experimental and theoretical results involving deep inelastic collisions with two- and three-body final states. In the last sections we will discuss some promising future aspects, i.e. collisions at intermediate energies and the possible use of electron-positron and electron-electron coincidences to extract information on the space-time evolution of the electronic states and maybe even to identify the spontaneous emission mechanism.

BASIC FORMALISM: DYNAMICS AND TIME-DELAY

The behavior of electrons and positrons during the collision of two very heavy nuclei is well described in a framework based on the adiabatic bound state interaction picture (Furry picture) of QED. In this approach, the second quantized electron-positron field operator $\hat{\Psi}(x)$ is expanded[5] in a complete set of solutions of the time-dependent independent-electron Dirac equation, which are in turn expanded into the instantaneous eigenfunctions of the two-center Dirac equation

$$[\vec{\alpha}\cdot\vec{p} + \beta m + V_1 + V_2 + V_{ee}]\varphi_k = E_k\varphi_k \tag{1}$$

where V_1, V_2 are the Coulomb potentials due to the two nuclei and V_{ee} is the effective electron screening potential. V_{ee} is usually calculated in the Hartree-Fock-Slater approximation for a highly (~50 times) ionized quasi-molecule,[6] assuming fully adiabatic motion of all electrons. The energy eigenvalues E_k depend parametrically on the nuclear separation R. The solutions of the time-dependent problem are found by writing

$$\phi_i(t) = \sum_j a_{ij}(t)\varphi_j(R(t))\exp(-i\chi_j(t)) \tag{2a}$$

with phase

$$\chi_j(t) = \int dt\, E_j(R(t)) \ , \tag{2b}$$

yielding coupled equations

$$\dot{a}_{ij} = - \sum_{k\neq j} a_{ik} \langle\varphi_j|\dot{R}\frac{\partial}{\partial R}|\varphi_k\rangle \exp(i\chi_j - i\chi_k) \tag{3}$$

for the occupation amplitudes a_{ij} with inital condition $a_{ij}(-\infty)=\delta_{ij}$. The matrix elements are asymptotically corrected for translational motion. In heavy collision systems with $(Z_1+Z_2)>137$, the $\partial/\partial R$-matrix elements are highly localized at small R, where the potential $(V_1+V_2+V_{ee})$ is dominated by its monopole component, and the wavefunctions retain approximately good angular momentum. Solutions of the full two-center problem, which are available, show that corrections to this monopole approximation are small for $s_{1/2}$ and $p_{1/2}$ ($\varkappa=\pm1$) states, even those in the continuum. (At the moment, the two-center states[7] are numerically too expensive to be used in the full problem. However, some persistent problems, e.g. a slight difference in beam energy dependence when theoretical and experimental results are compared, might have their origin in this approximation.)

Eqs.(3) have to be modified in the case of supercritical

fields, when the $1s_{1/2}$ state joins the lower Dirac continuum and becomes a resonance. This is done by a now well-established procedure borrowed from nuclear theory, in which the resonant bound state is truncated at the classical turning point and a new orthogonal continuum is constructed to complete the basis set.[8] Because these states are no longer eigenstates of the adiabatic Hamiltonian (1), additional potential matrix elements appear in eq. (3) which describe spontaneous positron emission. Although the truncation procedure is not unique, the resulting positron and electron spectra have been shown to be invariant.

Many-electron effects within this independent particle approximation are easily taken into account by appropriate summation of amplitudes over single-particle states. When the initially occupied states are denoted by a Fermi surface F, the final number of excited particles in state k is:

$$N_k = \sum_{i<F} |a_{ik}|^2 \quad , \quad (k>F) \tag{4a}$$

the number of holes, or positrons, in state k is

$$\bar{N}_k = \sum_{i>F} |a_{ik}|^2 \quad , \quad (k<F) \tag{4b}$$

and the number of particle hole pairs is described by an accidental and a correlation term:

$$N_{ke} = N_k \bar{N}_e - \sum_{\substack{i<F \\ j>F}} a^*_{ik} \, a_{ie} \, a^*_{je} \, a_{jk} \; . \tag{5}$$

In the case of electron-positron coincidences, also angular correlations occur. If only $s_{1/2}$ and $p_{1/2}$-states are taken into account, which should be a good approximation if the electron energy is not too low, the expression is

$$\begin{aligned} N_{\vec{p}_e,\vec{p}_p} &= \sum_{\lambda_e,\lambda_p} N_{\vec{p}_e\lambda_e,\vec{p}_p\lambda_p} = (4\pi)^{-2}\Big\{4(N_{E_e}^{(-)} + N_{E_e}^{(+)})(N_{E_p}^{(-)} + N_{E_p}^{(+)}) \\ &+ 2\left|\sum_{r<F} a_{r,E_e}^{(-)} a_{r,E_p}^{(-)*}\right|^2 + 2\left|\sum_{r<F} a_{r,E_e}^{(+)} a_{r,E_p}^{(+)*}\right|^2 \\ &+ 4\cos\theta_{ep}\cdot \mathrm{Re}\left[\left(\sum_{r<F} a_{r,E_e}^{(-)*} a_{r,E_p}^{(-)}\right)\left(\sum_{r<F} a_{r,E_e}^{(+)} a_{r,E_p}^{(+)*}\right)\right]\Big\} \end{aligned} \tag{6}$$

A similar expression, except for signs, holds for electron-electron angular correlations:

$$\begin{aligned} N_{\vec{p}_i,\vec{p}_j} &= (4\pi)^{-2}\Big\{4(N_{E_i}^{(-)} + N_{E_i}^{(+)})(N_{E_j}^{(-)} + N_{E_j}^{(+)}) \\ &- \Big(2\left|\sum_{r<F} a_{r,E_i}^{(-)} a_{r,E_j}^{(-)*}\right|^2 + 2\left|\sum_{r<F} a_{r,E_i}^{(+)} a_{r,E_j}^{(+)*}\right|^2 \\ &+ 4\cos\theta_{ij}\cdot \mathrm{Re}\left[\left(\sum_{r<F} a_{r,E_i}^{(-)*} a_{r,E_j}^{(-)}\right)\left(\sum_{r<F} a_{r,E_i}^{(+)} a_{r,E_j}^{(+)*}\right)\right]\Big)\Big\} \end{aligned} \tag{7}$$

The general properties of QED of strong fields are now well understood, both from the theoretical and experimental points of view. The measured $e^{\pm}$ and e^{-}-spectra are structureless for collisions without nuclear contact, and agree very nicely with the calculated spectra. This is shown in Fig. 1 and 2 for electron spectra from Au+U-collisions and positron spectra measured in several collision systems involving a total nuclear charge $Z = Z_{\mathrm{Target}} + Z_{\mathrm{Projectil}}$ ranging from Z = 163 to Z = 188. The electron spectra are approximately exponential at energies above 200 keV, whereas the positron spectra exhibit a broad maximum around 400 keV due to Coulomb repulsion. The shape of these spectra can be understood in terms of simple scaling models based on the approximate 1/R-dependence of the coupling matrix elements, but their absolute value

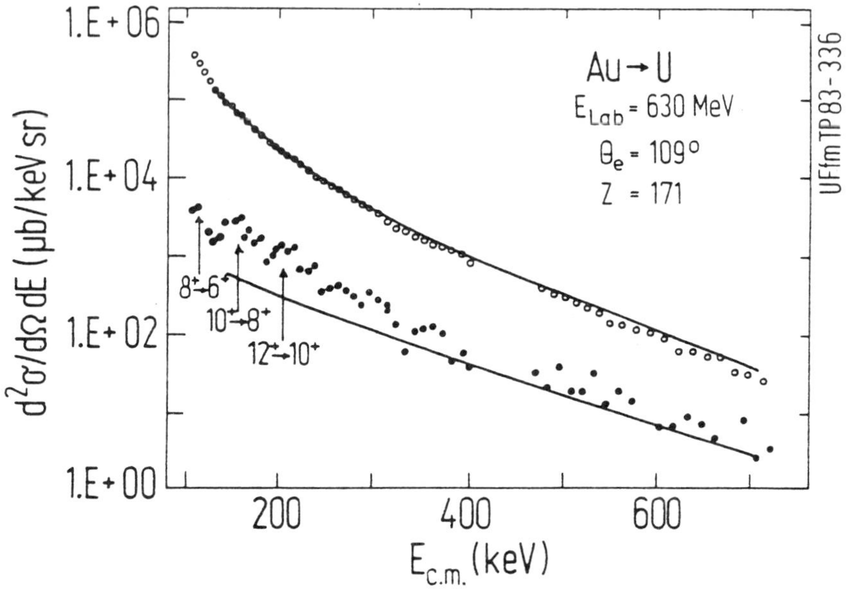

FIGURE 1. The energy spectrum δ-electrons emitted in Au+U collisions at 630 MeV Lab-energy measured by Koenig et al. in comparison with coupled channel calculations.[9] Upper curve: total electron yield, lower curve: Coincidence with a K-shell hole is required. Several nuclear conversion lines are visible.

requires full solutions of the coupled equations (3).

The essential ingredients of these equations (3) are the coupling-matrix-elements M_{ik} of the form

$$M_{ik} = \dot{R}\langle\varphi_i|\frac{\partial}{\partial R}|\varphi_k\rangle \qquad (8)$$

in subcritical collisions, where R denotes the two-center-separation of the colliding nuclei. In supercritical collisions, when a resonance develops in the spectrum of the adia-

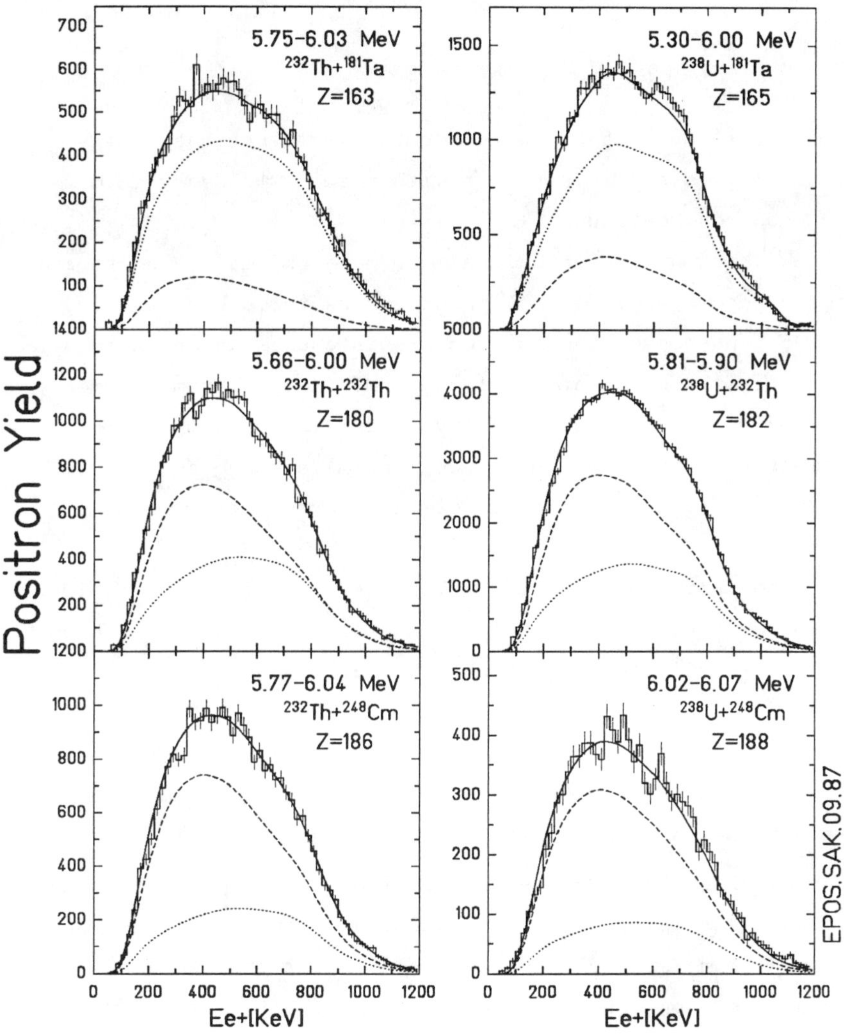

FIGURE 2. Positron spectra from several heavy collision systems as indicated measured by the EPOS group.[10] Dashed lines: Calculated atomic positrons, dotted lines: Nuclear conversion background, full lines: sum of both.

batic Hamiltonian, an additional coupling will enter the problem if the resonance projection method mentioned above is employed.

$$M_{ik} = \dot{R}\langle\varphi_i|\frac{\partial}{\partial R}|\varphi_k\rangle + i\langle\varphi_i|\hat{H}|\varphi_k\rangle \qquad (9)$$

The latter coupling accounts for the fact that the supercritical 1s-state is unstable with respect to decay into the positron continuum.

Working within an adiabatic expansion basis the original time dependence of the collision enters through a certain time dependence of the two-center-distance R which in turn parameterizes each matrix element $M_{ik}(R(t))$. From a numerical point of view this is very useful, since all matrix-elements (for a given collision system) are subject of computation only once on a sufficiently narrow one dimensional R-grid.

The time evolution of R=R(t) resulting from the nuclear dynamics of a specific collision determines the time dependence of the Dirac-field through the coupled-channel-equations (3).

For collisions below the Coulomb-barrier the use of classical Rutherford-trajectories reproduces the measured δ-electron- and positron-spectra very well (Fig. 1,2).

In order to describe deep inelastic nuclear scattering processes as well, the time evolution of the nuclear separation R, or more precisely the evolution of the nuclear charge distribution, needs to be altered. With presently achievable accuracy the experiments are sensitive on the dominant features of the inelastic process. The Q-value of the reaction and the multiplicity of the final state (sequential fission) can be used as triggers to select central or peripheral collisions.

For the theoretical analysis simple schematic parametrizations of the trajectory R(t) or various semiclassical friction models have been employed.[12] One such model applicable to heavy ion collisions at low energies (close to the Coulomb-barrier), is that of R. Schmidt et al.[13] which will be referred to in the subsequent discussion.

The nuclear delay-time T was found to be the most important property of the nuclear dynamics being detectable through electronic excitation spectra. A fixed value of the time-delay leads to characteristic oscillating structures in the spectra of emitted δ-electrons and positrons.[3]

Latest experiments on deep inelastic nuclear scatterings show certain discrepancies, when compared to theoretical calculations. In the collision system Pb+Pb at 8.6 MeV/u deviations by a factor 2 occur in the absolute values of differential δ-electron as well as positron production cross-sections. Fig. 3 shows the situation for the δ-electrons and indicates that the structures are less pronounced than predicted by a simple reaction model.

The emission spectra resulting from elastic nuclear collisions are better described by theory. This is illustrated in Fig. 4 involving the same collision-system, but the elastic channels only. Therefore there is a need for a more realistic description of deep inelastic nuclear processes in the calculations, in particular the inclusion of fluctuation phenomena.

Recently the theoretical formalism has been generalized to include the effect of sequential fission in the outgoing nuclear channel.[14] Such events occur due to energy dissipation into internal degrees of freedom of the colliding nuclei. A parameterized description of the fissioning into three

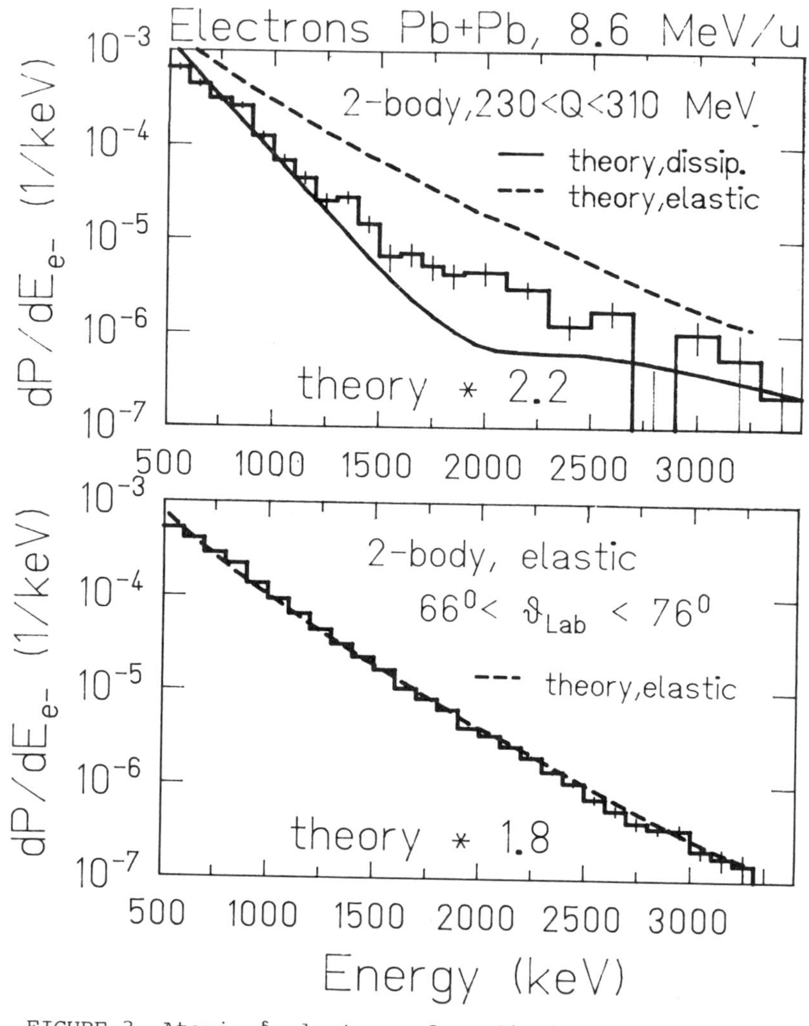

FIGURE 3. Atomic δ-electrons from dissipative binary Pb+Pb collisions[11] at 8.6 MeV/u Lab-energy, compared with theoretical results using nuclear trajectories from a friction model (full line).

FIGURE 4. Atomic δ-electrons from elastic Pb+Pb collisions at 8.6 MeV/u Lab-energy compared with theory.

separated charged centers, followed by the corresponding three-center-motion was entered as a new nuclear dynamics to the coupled channel formalism for the Dirac field.

Thus the nuclear configuration depends on further parameters, fixing the fission properties. Consequently the matrix-elements M_{ik} depend on several coordinates. Since tabulation of matrix elements in this multidimensional parameter space for sufficiently many values was prohibitive, they had to be recalculated for every single trajectory. The expense of computing time then is much larger compared to two-center calculations.

Fig. 5 depicts a comparison of two-center and three-center calculations for δ-electron emission in the monopole approximation, applied to the system Pb+U.

i) Assuming a deep inelastic incoming two-body channel with 8.4 MeV/u and impact parameter b=4.3 fm, the sequential fission of U is taken symmetric with a Q-value of 280 MeV. The fission process starts when the nuclei are 30 fm apart in the outgoing channel and lasts for a time interval of 150 fm/c.

ii) Assuming a central elastic collision with 6.0 MeV/u the parameters are the same, but the beginning of fission is even earlier at a two-center distance of 20 fm.

Deviations are nowhere larger than 14% and show no other structural properties than the two-body spectra. Thus we conclude that calculations may be restricted to the two-center case, while two-, three- as well as four-body events may be used experimentally for the analysis of nuclear reaction times. The situation may be different for bound-state vacancy production due to vacancy sharing processes occuring at very large nuclear separation.

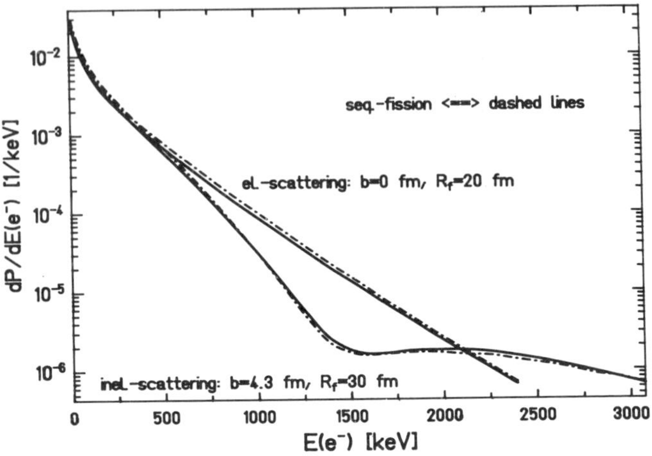

FIGURE 5. Comparison of theoretical δ-electron spectra from Pb+U collisions assuming binary and ternary nuclear exit-channels. Elastic scattering: 6.0 MeV/u impact energy with impact parameter b=o fm. Dissipative collision: 8.4 MeV/u impact energy with impact parameter b=4.3 fm.

ELECTRON EXCITATIONS AT INTERMEDIATE IMPACT ENERGIES

Heavy ion collisions at intermediate impact energies ($\lesssim$ 100 MeV) are currently receiving considerable interest since they probe the behaviour of nuclear matter, in a region between the low- and high energy limits. In low energy collisions electronic excitation processes as described in previous sections are established to be a useful tool to study the reaction dynamics. This holds true also for the range of intermediate energies. If the nuclear motion is still non-relativistic the formalism described above may be applied immediately.[15] At still higher collision energies the full retarded electromagnetic potential must be used and the question of convergence of the basis expansion has to be reconsidered. Even in the nonrelativistic region calculations for intermediate energy collisions[16] are more demanding.

i) Classical friction models, describing the nuclear motion for low impact energy (previous section) need to be altered.

ii) The numerically considered basis set $\{\varphi_i\}$ must be enlarged adequately into regions of higher electron energy, because excitation into such states is supposed to be most sensitive on nuclear dynamics.

iii) As a consequence of ii) the set of matrix-elements M_{ik} will include states of high energy, which exhibit oscillatory behaviour as a function of the two-center separation R. This calls for refined interpolation procedures.

Fig. 6 depicts the oscillations of various radial coupling matrix elements $M_{ik}(R)$ as a function of the two-center distance R. The matrix elements in fig. 6a describe pair creation-processes involving electronic and positronic states, whereas in fig. 6b continuum-continuum-couplings of electronic states only are shown.

The finite size of the nuclei under consideration effects couplings between states of high energy rather strong. This is illustrated in fig. 7. The matrix element in fig. 7a was calculated with respect to finite nuclear size, in contrast to the one in fig. 7b evaluated for point-like nuclei. The depicted matrix elements were obtained in monopole-approximation, electron screening was neglected.

In coupled-channel-calculations at energies close to the Coulomb barrier (cf. previous section), such structures in the couplings did not arise, since the excitation spectra, calculated up to approximately 2.5 MeV total electron energy

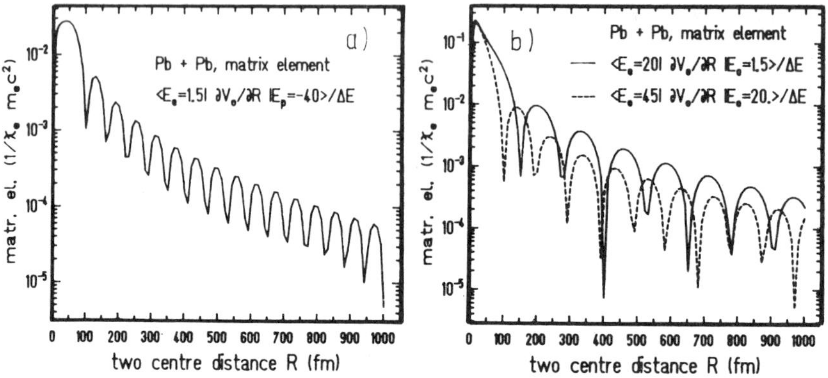

FIGURE 6. Radial coupling matrix elements for the system Pb+Pb: a) Coupling between a positronic and a electronic state, describing pair-creation processes. b) Continuum-continuum couplings of electronic states. The indicated energies correspond to the total energy in units of the electrons's rest mass.

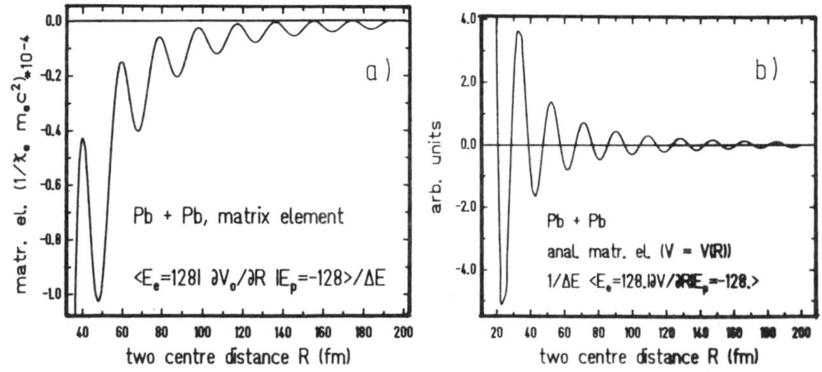

FIGURE 7. Radial couplings describing the creation of a pair at $E=\pm 128mc^2$.
a) The matrix element as a function of R considering a nuclear finite size.
b) The same matrix element, assuming pointlike nuclei. The absolute values should not be compared directly.

do not involve states that cause these structures.

Theoretical predictions for δ-electron emission in central Pb+Pb collisions at 60 MeV/u are shown in figure 8. The full (dashed) line indicate the contributions resulting from the ionization of $s_{1/2}\sigma$ ($p_{1/2}$) states.

There is no mixing when the monopole-approximation is applied so that the total emission rate is given by the sum over both contributions. The spectrum is dominated by electrons ionized out of the 1sσ-state. At electron energies $\lesssim$ 10 MeV oscillations are observed which arise if a binary nuclear exit channel is assumed in the employed model for the nuclear reaction with a time delay of $1\text{-}2\cdot 10^{-21}$ s.

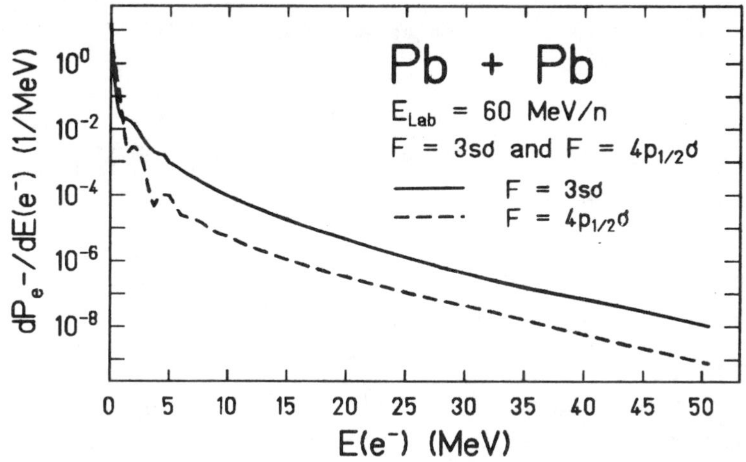

FIGURE 8. δ-electron emission in central collisions of Pb+Pb at an intermediate Lab-energy of 60 MeV/u. The contribution of $\varkappa=\pm 1$ channels is shown separately. F indicates the Fermi-level.

DETECTING SUPERCRITICAL FIELDS

i) Crossed beams of heavy ions

In the near future beams of bare U-nuclei will become available at the SIS-ESR storage ring. This promises the chance to do colliding beam experiments with fully stripped heavy ions moving in parallel direction, i.e. at small relative kinetic energy. Such experiments could be an improved tool to test the spontaneous decay of the QED vacuum in strong fields, as predicted several years ago. In present experiments the presence of electrons in the inner atomic shells to a large degree blocks positron creation relating to the "supercritical" quasimolecular 1sσ-state.

If one considers a collision of bare heavy nuclei,[17] the positron-production therefore will be enhanced strongly. The absence of electron screening adds to this effect. Fig. 9 gives an illustrative example for the change of positron - production when going from ordinary (9a) to bare (9b) U+U collisions. The full lines give the spectra of positrons emitted in elastic Rutherford scattering at $E_{C.M.} = 740$ MeV into an angular window of $\theta_{cm} = 90^{o} \pm 6^{o}$. Superimposed is the contribution of spontaneous positron emission calculated under the hypothetical assumption that there is a nuclear reaction component with the very long time delay $T = 5 \cdot 10^{-20}$ s and a cross section $\sigma_{nuc} \simeq 7$mb. The effect is an enhancement of differential production crossections by a factor 40 and an increased sensitiveity on the time delay. Since induced and spontaneous positron creation both are enhanced in a similar fashion by the absence of Pauli blocking however, bare collisions do not offer a simple qualitative experimental signal to disentangle these effects.

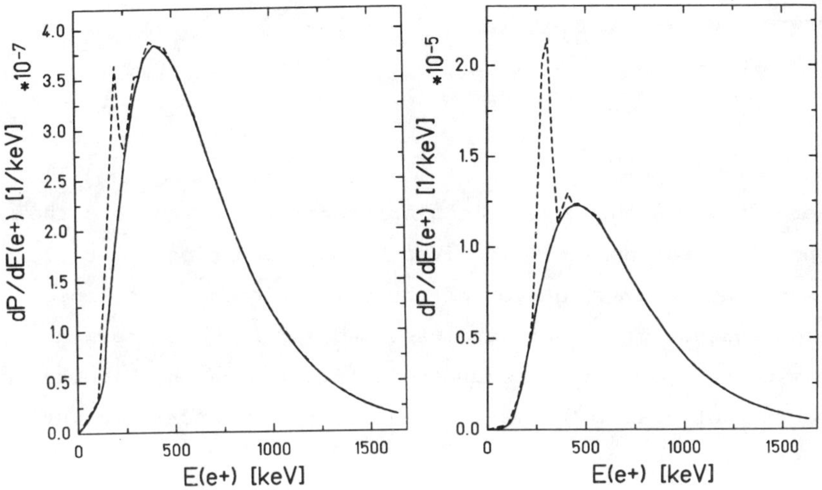

FIGURE 9. Positron creation spectra for U+U collisions of (a) on intermediately stripped U projectile impinging on an atomic U target, (b) completely stripped projectile and target ions. Full lines: Rutherford scattering, dashed lines: the Consequence if a fraction of collisions has a very long ($5\cdot10^{-20}$s) nuclear time delay.

ii) Quasimolecular interference effects

A rather model-independent method to study the binding energies of quasimolecular orbits can be based on phase interferences of MO X-ray radiation.[18,19] At high X-ray energies one observes electronic transitions into the 1sσ-orbital. Normally this leads to a broad continuum due to the variation of $E_{1s\sigma}(R(t))$ and to collision broadening. If a 1sσ-vacancy is brought into the collision, however, by using a projectile which is fully stripped (or has only one electron left, e.g. U^{91+}) the situation is changed. X-rays of the same frequency can be emitted in the ingoing and outgoing branch of the collision which leads to a characteristic interference pattern of the X-ray spectrum. Roughly speaking it is determined by

$$\frac{d\sigma_x}{dE_x} \sim \cos\left(\int_{R_{min}}^{R(E_x)} dR \; \frac{\Delta E(R)}{\dot{R}} \right) \qquad (10)$$

where $\Delta E(R)$ is the quasimolecular transition energy and the upper integration limit is given by the condition $E_x = \Delta E(R)$. A systematic investigation of these interference patterns in dependence of impact parameter and collision energy can be expected to lead to accurate information on the binding energies of quasimolecular states also in supercritical systems.

iii) Angular correlation of e^+-e^-- pairs

For the following discussion[20] we return to equation (6), which describes the number of coincident electron-positron pairs created in a collision in dependence of their relative angle θ_{ep}. We identify the first term as the contribution of accidental correlations the third and the second one as a contribution of true correlations with- and without angular dependence. For convenience we write equation (6) in the form

$$P\left(\theta_{ep}\right) = a_o + b_o \cdot \cos\left(\theta_{ep}\right) + A, \qquad (11)$$

A stands for accidental correlations. According to eq. (6) the anisotropy coefficient b_o depends on the interference between $\varkappa = -1$ and $\varkappa = +1$ amplitudes. Thus it will be particularly sensitive to effects which distinguish between these two channels, e.g. the spontaneous positron production process.

Unfortunately the coefficient b_o under ordinary circumstances is quite small compared with the accidental part A. By going to collisions of bare nuclei A can be nearly fully switched off, since here no δ-electrons can be ionized out of bound atomic states. In order to study the characteristic properties of the coefficients a_o, b_o we compare the results from a supercritical- and subcritical system respectively. Collisions with modest time delay have been studied within a schematic model. The nuclear configuration is treated to be frozen at the distance of closest approach for a certain delay time T. This is applied to the systems U+U and Pb+Pb at 6.2 MeV/u impact energy and impact parameter b=0fm.

Fig. 10 represents the anisotropic coefficient b_o integrated over a double energy-window 102 keV $< E_{e^-}, E_{e^+} <$ 588 keV to give the corresponding pair emission rate as a function

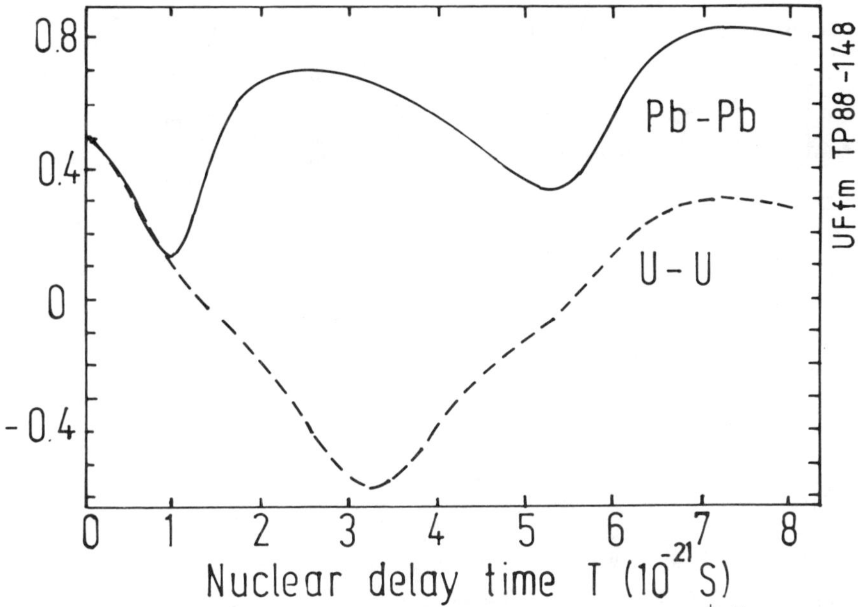

FIGURE 10. The anisotropic coefficient b_o for e^+e^- pair emission, integrated over a double energy-window 102 keV$<$ $E_{e^-}, E_{e^+} <$ 588 keV as a function of the nuclear time delay T. The curves correspond to supercritical (dashed line) collisions of U+U and subcritical (full line) collisions of Pb+Pb at 6.2 MeV/n and b=o fm.

of delay time T. Note that the value of b_o may becomes negative in the supercritical system (dashed line) whereas it remains positive throughout in the other case (full line).

A coefficient $b_o > o$ corresponds to enhanced parallel e^+e^- emission in contrast to $b_o < o$ which favours an antiparallel emission of electron and positron.

The effect is further illustrated in figs. 11a, b calculated under the same conditions as before except for the assumption that there is a Gaussian distribution for the delay time T in one case, and vanishing delay time in the other case. The plotted properties are now correlated - and angular correlated pair emissions a_o, b_o referring to an electron energy of 332 keV, as a function of the positron energy. The sign of $b_o(E_{e^+})$ provides again a qualitative difference comparing the sub- and the supercritical system that allows b_o to be negative in the latter case.

SUMMARY

Experiment and theory in QED of strong fields in heavy ion collisions are in good general agreement. The influence of nuclear time-delays on δ-electron spectra has already allowed for the determination of reaction times in the order of 10^{-21} s. Future prospects exist for a rather detailed measurement of the collision dynamics, including the extent of fluctuations of the delay time. At higher collision energies the electron spectra will extend to energies in the tens of MeV, giving access to nuclear reaction times on a scale of 10^{-23} s. If this prospect can be realized experimentally, much improved computational facilities are required to perform quantitative calculations in the coupled channel for realistic nuclear reaction models. On the low-energy side, new

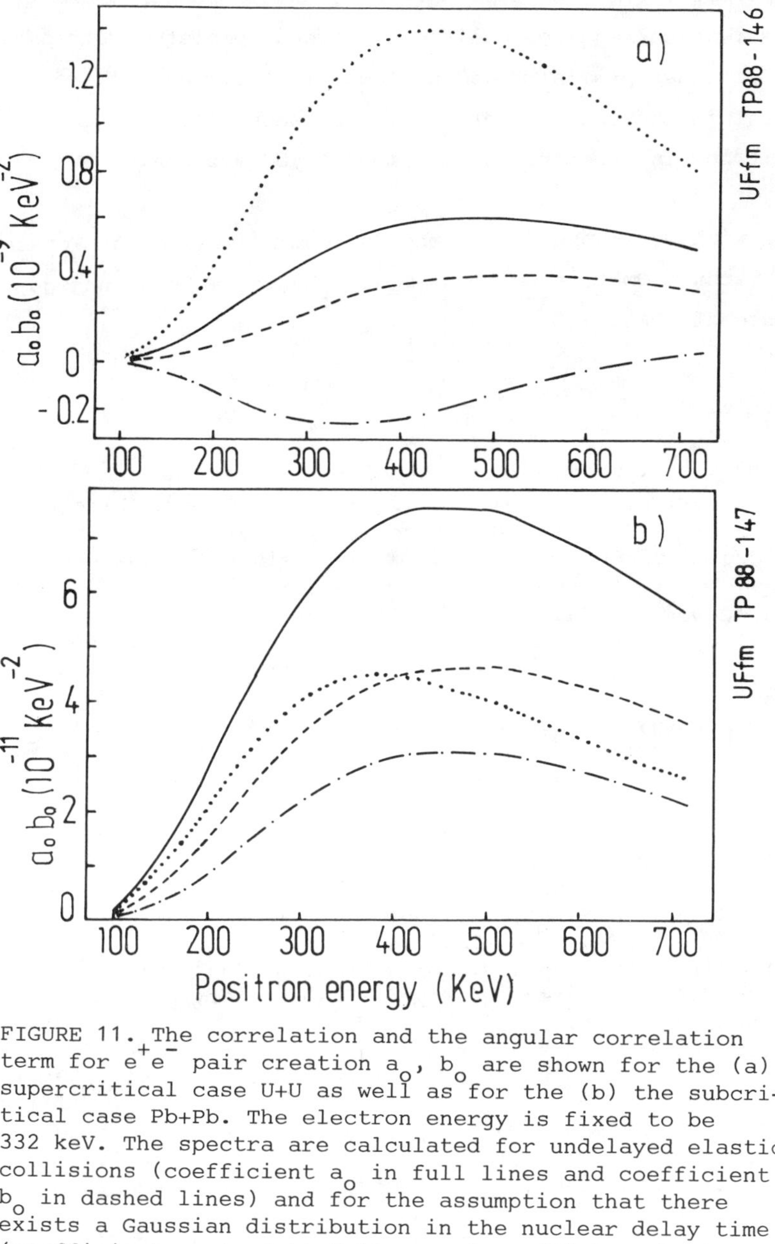

FIGURE 11. The correlation and the angular correlation term for e^+e^- pair creation a_o, b_o are shown for the (a) supercritical case U+U as well as for the (b) the subcritical case Pb+Pb. The electron energy is fixed to be 332 keV. The spectra are calculated for undelayed elastic collisions (coefficient a_o in full lines and coefficient b_o in dashed lines) and for the assumption that there exists a Gaussian distribution in the nuclear delay time T (coefficient a_o in dotted lines and coefficient b_o in dashed-dotted lines). The properties of this distribution are $\bar{T} = 2 \cdot 10^{-21}$ s and $\tau = 7 \cdot 10^{-22}$s.

proposals for the detection of spontaneous positron creation exist, which require collisions between completely stripped nuclei. Most promising among these is the predicted reversal in the angular correlation between electrons and positrons in supercritical, time-delayed systems.

This work was supported by the Bundesministerium für Forschung und Technik BMFT, as well as by the Deutsche Forschungsgemeinschaft DFG.

REFERENCES

1. For a review and collection of references see W. Greiner, B. Müller and J. Rafelski, Quantum Electrodynamics of Strong Fields (Springer, Heidelberg, 1985). Physics of Strong Fields, ed. W. Greiner, Plenum Press, New York, (1987).
2. T. Cowan, H. Backe, K. Bethge, H. Bokemeyer, H. Folger, J. S. Greenberg, K. Sakaguchi, D. Schwalm, J. Schweppe, K. E. Stiebing, P. Vincent, Phys. Rev. Lett. 56, 444 (1986).
   W. Koenig, F. Bosch, P. Kienle, C. Kozhuharov, H. Tsertos, E. Berdermann, S. Huchler, W. Wagner, Positron Lines from subcritical heavy ion-atom collisions, Z. Phys. A328, 129 (1987).
3. G. Soff, J. Reinhardt, B. Müller, W. Greiner, Phys. Rev. Lett. 43, 1981 (1979).
4. H. Backe, P. Senger, W. Bonin, E. Kankeleit, M. Krämer, R. Krieg, V. Metag, N. Trautmann, J. B. Wilhelmy, Phys. Rev. Lett. 50, 1838 (1983).
   R. Krieg, E. Bozek, U. Gollerthan, E. Kankeleit, G. Klotz-Engmann, M. Krämer, U. Meyer, H. Oeschler, P. Senger, Phys. Rev. C34, 562 (1986).
5. For an alternative approach see C. Böttcher, M. R. Strayer, Phys. Rev. Lett. 54, 669 (1985).
6. T. H. J. de Reus, J. Reinhardt, B. Müller, W. Greiner, G. Soff, and U. Müller, J. Phys. B17, 615 (1984).
7. K. H. Wietschorke, P. Schlüter, G. Soff, K. Rumrich, W. Greiner, Phys. Rev. A36, 377 (1987).
8. J. Reinhardt, B. Müller, W. Greiner, Phys. Rev. A24, 103 (1981).

9. G. Mehler, T. de Reus, J. Reinhardt, G. Soff, U. Müller, Z. Phys. A320, 355 (1985).
M. A. Herath-Banda, A. V. Ramayya, C. F. Maguire, F. Güttner, W. Koenig, B. Martin, B. Povh, H. Skapa, J. Soltani, Phys. Rev. A29, 2429 (1984).
10. K. Sakaguchi, H. Backe, K. Bethge, H. Bokemeyer, T. Cowan, H. Folger, J. S. Greenberg, P. Salabura, D. Schwalm, J. Schweppe, and K. E. Stiebing, to be published.
11. M. Krämer, B. Blank, E. Bozek, E. Kankeleit, G. Klotz-Engmann, C. Müntz, H. Oeschler, M. Rhein, IKDA 87/36, preprint (1987).
12. U. Müller, J. Reinhardt, G. Soff, B. Müller, and W. Greiner, Z. Physik A297, 357 (1980).
U. Müller, G. Soff, J. Reinhardt, T. de Reus, B. Müller, W. Greiner, Phys. Rev. C30, 1199 (1984).
13. R. Schmidt, V. D. Toneev, G. Wolschin, Nucl. Phys. A311, 247 (1978).
14. St. Graf, J. Reinhardt, U. Müller, T. de Reus, B. Müller, W. Greiner, G. Soff, Z. Phys. A329, 365 (1988).
15. T. de Reus, J. Reinhardt, B. Müller, W. Greiner, G. Soff, Phys. Lett. 169B, 139 (1986).
16. T. de Reus, Ph D thesis, Frankfurt (1987).
17. U. Müller, T. de Reus, J. Reinhardt, B. Müller, W. Greiner, G. Soff, Phys. Rev. A37, 1449 (1988).
18. R. Schuch, H. Schmidt-Böcking, I. Tserruya, B. M. Johnson, K. W. Jones, M. Meron, Z. Physik A320, 185 (1985).
19. J. Kirsch, B. Müller, W. Greiner, Z. Phys. A1, 47 (1985).
20. O. Graf, S. Reinhardt, T. de Reus, B. Müller, W. Greiner, G. Soff, Z. Phys. A328, 81 (1987).
O. Graf, J. Reinhardt, B. Müller, W. Greiner, G. Soff, GSI-88-22 Preprint (1988).

PRECISION CALCULATIONS OF QED EFFECTS IN ATOMS

J. SAPIRSTEIN
University of Notre Dame, Notre Dame, Indiana, U.S.A.

Abstract Calculations in QED that require supercomputer power are discussed. The use of many-body perturbation theory in making predictions of the properties of many-electron atoms is described, and the study of high-Z few-electron ions advocated as a method of studying QED radiative corrections in intense Coulomb fields.

The rapid increase in computational power that is becoming available to theoretical physicists is having a profound effect on a great number of disciplines. I would like to discuss today the impact that can be expected on one of the most successful branches of physics, Quantum Electrodynamics (QED). The main point that I would like to make is that there are two aspects of QED that can benefit from this great increase in our computational abilities. The first has to do with 'classic' QED, that is its applications to simple atomic or electronic systems, specifically the one-electron atoms, including hydrogen-like atoms and the exotic atoms muonium and positronium, and the leptonic anomalous magnetic moments. In these systems, there is present enough simplicity so that relatively small field theoretic effects can be studied to finer and finer levels of precision. The computational problem posed by the study of these systems has to do with the behavior of radiative corrections, which in general lead to complicated many-dimensional integrals that defy analytic evaluation except in certain special cases. At this point a very specific demand is placed on computing power: one needs to evaluate lengthy integrands at as large a number of points as is possible: this computing task is conceptually quite simple, but can lead to problems of accuracy that strain modern supercomputers, as will be discussed in the section on the calculation of anomalous magnetic moments.

The other computational demand associated with QED is more difficult to quantify, because I do not know precisely the necessary calculations required. It is perhaps better phrased as the possible opening up of a new field that may be made possible by the explosion of computer power becoming available. It is simply the extension of this extremely successful theory, which routinely makes predictions on the systems mentioned above that agree with experiment at the six to seven digit level, to the rest of the periodic table and to molecules. The accuracy with which the theory works in simple systems makes it a rather safe assumption that QED is the underlying theory of atoms and molecules. However, because once more than one electron is present, with the sole exception of helium, there is no practical method I know of that can routinely provide energy levels or other physical properties of these systems accurate to the five or six digit level necessary to discern the characteristic field theory effects of QED from uncertainties of atomic structure. The problem is not one that can be resolved by simply imagining a direct numerical solution of a many-particle Schrödinger or Dirac equation made possible by future supercomputers. The difficulty is that so many variables are present in these equations for atoms with more than three or four electrons that even a thousandfold increase in computer power cannot be expected to lead to accurate solutions. Some other method may, however work: the most intensive efforts in finding such methods have been made in the field of quantum chemistry. In particular, I would like to discuss the use of a tool used by that community, along with solid state and nuclear physics, many-body perturbation theory (MBPT)[1]. By MBPT I mean to include direct perturbative studies along with the large variety of methods that have been devised to account for infinite subsets of contributions, such as coupled cluster methods. The reason for concentrating on this technique, outside of the fact that it is rather successful, particularly in application to the study of the spectra of high-Z few-electron ions, is that it is a method that fits into QED, in the sense that there are well-defined QED calculations that contain within them the results of MBPT. In addition, however, the QED calculations also have the characteristic field theoretic effects of negative energy states, photon retardation, and the need for ultraviolet renormalization. Thus the second topic I wish to discuss is the kind of calculation one might imagine possible sometime in the next thirty years: the ab-initio treatment of a complex atom or simple molecule with such completeness that agreement with experiment cannot be found until QED effects are added in. Fortunately, because of the highly ionized systems mentioned above, a glimpse of this future can be had right now, and I will discuss the application of QED and MBPT to these systems.

Without doubt the most striking test of QED is afforded by the study of the electron anomalous magnetic moment. This results as a confluence of the development on the experimental side of trapping techniques[2], which enable one to deal with a single electron in the presence of relatively simple fields, and on the theoretical side of the large scale calculations carried out over the last 15 years by Kinoshita and collaborators[3], which calculations are possible only because of supercomputers. There is actually a limitation on the experimental side of the situation that is related to the effect of the trap itself on the motion of the electron that has been treated by Brown et. al.[4] The problem is that a careful analysis of the motion of the electron leads to mode structure that is very difficult to control: because of this it will be difficult to push the experimental precision under one part in 10^{12}. Fortunately, this precision is so high that there is a great deal of room for testing the higher order QED corrections, which now have been extended to fourth order in the fine structure constant α. The theoretical prediction can be written as

$$g-2 = \sum_{n=1}^{\infty} a_n \left(\frac{\alpha}{\pi}\right)^n \quad , \tag{1}$$

where the first two constants a_n are known analytically,

$$a_1 = \frac{1}{2} \tag{2a}$$

$$a_2 = \frac{197}{144} + \frac{\pi^2}{12} + \frac{3}{4}\zeta(3) - \frac{1}{2}\pi^2 \ln 2 \tag{2b}$$

and the next two numerically[3] (although it should be pointed out that the great bulk of contributions to a_3 have been evaluated analytically),

$$a_3 = 1.1765(13) \tag{3a}$$

$$a_4 = -.8(1.4) \tag{3b}$$

The evaluation of a_4 is one of the most striking examples of a calculation of a fundamental physical quantity that would be completely impossible without supercomputers. The first difficulty with the calculation is one simply of logistics. The chief characteristic of perturbation theory is that it becomes more and more unwieldly as higher and higher orders are considered. In the present case, there are 891 Feynman diagrams that need to be considered. Even if each were trivial, the simple organization of so many terms is a non-trivial problem. A major feature of the calculation is a skillful use of a Ward-like identity to group together numbers of vertex graphs with a common origin from a self energy graph. This enables one to concentrate on a distinctly smaller number of self-energy graphs, which number 'only' 47 for the heart of the calculation, those terms involving no closed electron loops: most of the other terms can be grouped with the well-understood three-loop calculation. These graphs can be converted into integrals over a ten-dimensional hypercube. While in lower dimensions there are a large variety of techniques to evaluate multi-dimensional integrals, the best method for high numbers of dimensions is adaptive Monte-Carlo integration[5], in which the function is evaluated at a large number of random points, which are however chosen to concentrate in regions where the function is either large or rapidly varying. When working properly, such techniques have an error which falls as $1/\sqrt{N}$, where N is the number of random points. For this reason, such integration methods will always easily be able to saturate any supercomputer: if a computer two orders of magnitude faster becomes available, one will simply run the same program to get a factor of ten smaller error. This particular application is complicated by the presence of the order of ten thousand terms for each graph coming from the complex Dirac numerator structure, and by the fact that each graph as it stands is divergent both in the ultraviolet and the infrared. While the renormalization of QED is in principle well understood, its actual implementation at this high order is quite delicate. There are usually several layers of subtraction required to render each integral finite, and in some cases the cancellations are so severe that the use of quadruple precision is required. One final difficulty with this calculation, typical of QED, is that there are cancellations of more than an order of magnitude between different graphs: this is the reason for the large error in a_4: very large scale calculations by Kinoshita have recently significantly reduced that error[6].

While extremely demanding of computer time, one advantage of this kind of calculation is that it is easily vectorized. The function must be evaluated at a large number of points, but these points are independent of one another. Thus evaluations at a number of different points can be carried out simultaneously, which favors machines with long vector lengths. In particular, implementation of these programs on the CYBER 205 provides dramatic increases in speed over scalar mode. An important feature of Monte-Carlo integration programs is their complete generality: they are of use in any field in which multi-dimensional integrals arise. In particular, they can be used in the great bulk of QED calculations, and not just for the electron anomaly. I leave the subject of the classic tests of QED with the observation that at present there is good agreement between theory and experiment in a wide range of systems, but that in order to advance the theoretical predictions large-scale computations are necessary that will be greatly facilitated by supercomputers.

The second main area in which supercomputing power is impacting QED is in the application of QED to many-electron atoms, which is closely related to the tool of many-body perturbation theory (MBPT). While the problem of understanding the structure of a general many-electron atom is of course a fundamental one of very basic interest, elementary particle physics is presently providing a powerful motivation for the development of techniques to solve this problem. This is because parity nonconserving (PNC) transitions have been measured in several kinds of heavy atoms, and in the case of cesium the experiments are achieving accuracies at the percent level[7]. In order to get accurate information about the nature of the PNC coupling, which depends on a quantity known as the weak charge Q_W, theory must be able to reliably predict the PNC transition rate for a given Q_W. However, while an alkali, cesium with its 55 electrons is an extremely complex system. If a successful calculational scheme can be set up, the benefits are high. While the Weinberg-Salam model[8] has been very successful in describing the weak interactions, it is important to test the model with ever-increasing precision. The theory was constructed to be renormalizable, so that loop effects, such as the electron anomaly discussed above, are calculable and finite. However, these effects typically come in at the percent level, so to test them that kind of accuracy is required of experiment and theory. Any deviation from the predictions could be the first sign of new physics, such as the presence of additional Z_o bosons (to which atomic PNC studies are particularly sensitive) or unexpected behavior associated with the relatively poorly understood Higgs sector. Even if the loop effects behave as expected, one still gets a determination of the Weinberg angle from any such experiment. There are a large number of high-energy physics experiments that are presently studying these questions. What is unique about the atomic physics effort is that the experiment is rather small scale, being a true table-top experiment. What I will argue next is that supercomputing power will be vital to its theoretical interpretation, but that if successful, information about unified theories of the weak and electromagnetic interactions just as valuable as that coming from very large scale accelerator experiments will be obtained.

Given this strong motivation to find techniques to study many-electron atomic structure with high accuracy, the question of what methods are most promising arises. The starting point of any calculation is the Schrödinger or Dirac Hamiltonian, which latter is

$$H_{DC} = \sum_{i=1}^{N} (\vec{\alpha}_i \cdot \vec{p}_i + \beta_i m - \frac{Z\alpha}{r_i}) + \alpha \sum_{i<j} \frac{1}{|\vec{r} - \vec{r}\,'|} \quad (4)$$

Because PNC is an effect which arises from very small distances, inside the nucleus, and because one is treating atoms with high nuclear charge, relativistic effects are significant, so the Dirac Hamiltonian is the best starting point. There are a variety of methods to calculate with this Hamiltonian: MBPT, multiconfigurational Dirac-Fock codes[9], or configuration-interaction (CI)[10] methods. However, there is a fundamental problem with the use of H_{DC}: while written down as a straightforward generalization of the many-electron Schrödinger Hamiltonian, it is not justified in any fundamental sense. The problem, which has been discussed by Sucher[11] and named 'continuum dissolution' has to do with negative energy states: H_{DC} does not 'know' that there is a filled negative energy sea, and consequently allows transitions into the negative energy continuum. One way to solve this problem is to introduce positive energy projection operators surrounding the second term of H_{DC}, which is the term responsible for these unphysical transitions. There is a solution however, in the framework of MBPT, that is extremely simple. As in any perturbation theory, MBPT involves intermediate sums over states, which for the relativistic case involve states of negative energy. There is a method of developing an analogous perturbation theory in QED, based on the Furry[12], or bound interaction representation, that leads to expressions almost identical to MBPT, along with characteristic field theoretic quantities such as the Lamb shift. However, because the filled negative energy sea is built into MBPT, the QED expressions do not always sum over negative energy states. At this point, unless one is doing a full QED calculation, one can simply use the QED framework to provide a set of rules about summing over negative energy states in a MBPT calculation to ensure the validity of the procedure, and then go on to our next topic, the investigation of MBPT in higher and higher orders.

For atoms with more than two electrons, the direct solution of the Schrödinger equation is basically impossible because of the explosion of variables in the wave function: for example, the spatial part of the cesium wave function involves 162 coordinates. MBPT solves this problem by having in lowest order each electron move independently in a spherically symmetric potential U(r), introduced so that

$$H_{DC} = H_0 + V_C,$$

$$H_0 = \sum_{i=1}^{N} (\vec{a}_i \vec{p}_i + \beta_i m + U(r_i)); \qquad (5)$$

$$V_C = \alpha \sum_{i<j} \frac{1}{|\vec{r}_i - \vec{r}_i'|} - \sum_{i=1}^{N} (U(r_i) + \frac{Z\alpha}{r_i})$$

The solution of $H_0|\varphi_0\rangle = E_0|\varphi_0\rangle$ is now trivial, since all the complication of the electron's mutual repulsion has been shifted to V_C, which can now be studied with standard Rayleigh-Schrödinger perturbation theory.

Let us now examine a particular term in the perturbation expansion for the transition matrix element of an alkali atom, which in lowest order would be

$$z_{wv} = \int d^3r \; z \; \psi_w^\dagger(\vec{r})\psi_v(\vec{r}) \tag{6}$$

where w and v are two valence states. An important term from second order MBPT that corrects this lowest order result is

$$z_{wv}^{(2)} = \sum_{amni} \frac{g_{aimn} \; z_{wi} \; g_{mnav}}{(\varepsilon_j - \varepsilon_v)\;(\varepsilon_m + \varepsilon_n - \varepsilon_a - \varepsilon_v)} \tag{7}$$

where the summation over a is over all core states, m and n over all excited states, i over both core and excited states with the exception of the valence state v, and the Coulomb matrix element g is defined as

$$g_{ijkl} = \int \frac{d^3r \; d^3r'}{|\vec{r} - \vec{r}'|} \; \psi_i^\dagger(\vec{r})\psi_k(\vec{r})\psi_j^\dagger(\vec{r}')\psi_l(\vec{r}') \; . \tag{8}$$

Because the states are formed in a spherically symmetric potential, the angle integrations and sums over the magnetic quantum numbers can be performed after partial wave decompositions are made: in this case this leads to

$$z_{wv}^{(2)} = \sum_{L=0}^{\infty} \sum_{[anmi]} \frac{C_L^2(am)C_L^2(nv)}{(2L+1)(2j_v+1)} \; \frac{r_{wi}R_L(aim\,n)R_L(mnav)}{(\varepsilon_i - \varepsilon_v)\;(\varepsilon_n + \varepsilon_m - \varepsilon_a - \varepsilon_v)} \tag{9}$$

where C_L is a reduced matrix element, and the radial integrals R_L are defined as

$$R_L(ijkl) = \int dr \int dr' \; \frac{r_<^L}{r_>^{L+1}} (g_i(r)g_k(r) + f_i(r)f_k(r)) \cdot (g_j(r')g_k(r') + f_j(r')f_k(r')) \tag{10}$$

where g is the upper radial component and f the lower radial component of ψ.

As it stands, this expression is afflicted with a number of infinite summations. In practice, the partial wave expansion converges fairly rapidly, with 80-90% of the answer usually coming from the partial waves L=0, 1, and 2. Past that point the convergence slows down, and for high accuracy it is desirable to go up to L=9 and to in addition perform some sort of extrapolation to estimate the remainder. At this point, however supercomputing power begins to play an important role. The problem is that as L increases, the number of angular momentum channels can increase very rapidly. A channel refers to a given set of values for κ_a, κ_m, κ_n, and κ_i in Eqn. 9. While these are restricted by the C_L factors, as L increases the number of allowed values can quickly run into very large numbers: this numbers become larger as higher orders of MBPT are considered.

The other infinite summations involve the sums over the excited states n,m, and i: a involves of course a finite sum, though for cesium it involves seventeen terms. These sums are actually the sum over the infinite set of discrete bound states together with an integration over the continuum of scattering states. A method of handling this problem that makes the evaluation of expressions such as Eqn. 9 particularly straightforward to code is the use of finite basis sets[13]. It is possible to construct a basis set of, say, 40 functions, with the first 10 forming the actual bound states of the problem, and the next 30 providing a representation of the continuum, such that performing the sums over these states give results accurate to more than 5 digits. Because of this, any excited state summation can be treated as a DO loop, and the coding of MBPT becomes quite automatic. However, if one uses much less than 40 states, the numerical accuracy of this method begins to deteriorate. Therefore, supercomputer power again becomes relevant: Eqn. 9 involves, for each angular momentum channel, a threefold summation that is in principle a triply nested DO loop that involve N^3 operations, where N is the size of the basis set. (Actually, in this case the i summation can be done independently, so the dependence is only N^2). As higher orders of perturbation theory are probed, this power gets higher and higher: for example, the fourth order energy involves a sixfold summation, with the number of channels for L=2 already in the thousands. Clearly supercomputer power is vital for systematic higher order MBPT studies of atomic properties. Given that MBPT provides a completely systematic approach, the next question is naturally how well it works. In particular, it would be desirable to be able to say which terms in MBPT need be evaluated in order to predict properties of cesium at the 1% level, so that the amount of computing necessary could be estimated. Unfortunately, there are no small parameters in this perturbation expansion as in a standard QED calculation: each term is nominally of the same order. Thus the best that can be done is to examine the behavior of the expansion 'experimentally'. In this section two such investigations into the behavior of hyperfine splitting and valence energy levels of cesium will be presented, with encouraging but somewhat inconclusive results. In the final section, however, we will discuss highly ionized few-electron ions. In this case a small parameter does exist, 1/Z: the extremely large nuclear Coulomb field in such ions makes the approximation of ignoring the inter-electron repulsion quite accurate; when that interaction is treated perturbatively, each higher order of MBPT is accompanied by another power of this factor. The convergence of the perturbation expansion is so rapid that wave function uncertainties are reduced far below the size of field-theoretic effects such as the Lamb shift. Before continuing, it should be mentioned that coupled cluster[1] techniques that sum selected classes of MBPT, do offer the hope of a small parameter for the neutral case. This parameter is the number of excitations out of the core: there is considerable evidence that restricting this number to be two or less picks up the dominant part of the perturbation expansion, with third and higher-order excitations giving small results. In the following only order-by-order MBPT will be discussed.

The ground-state hyperfine splitting of cesium is of course an experimentally well-known quantity, forming as it does the standard of time. Because it arises from the interaction of the electron spin with the magnetic moment of the nucleus, which emphasizes the short distance behavior of the wave function, it is similar to the PNC problem, and should be accurately predicted by any theoretical method used to calculate PNC effects. Unfortunately, in lowest order the most common choice for the central potential U, the frozen-core Hartree-Fock model, underestimates this quantity by 37%, as seen from Table 1.

TABLE 1 MBPT expansion of cesium 6s hyperfine constant

$A^{(0)}$	$A^{(1)}$	$A^{(2)}$	Sum	Expt.
1435 MHz	289MHz	642MHz	2366MHz	2298MHz

The first order corrections to this quantity can be grouped together with an infinite subset of terms associated with the RPA (Random Phase Approximation), and improve the agreement with experiment only slightly. More important are the second order corrections discussed above, which lead to an overestimate of the splitting by 2.8%. One's first thought might be that better convergence might be found with another potential that gives better lowest order answers. Our group has found, however, that although potentials can be found that give excellent lowest order results, that higher order perturbations are large and uncontrollable. This same state of affairs exists for several other cesium matrix elements, with poor lowest order results improving with the addition of the corrections mentioned above to the few percent level, with the worst disagreement being the 7s hyperfine splitting, which disagrees with experiment by 4.8%. This same sort of calculation has been extended to the PNC case[14], and gives a result for the Weinberg angle consistent with the present world average. However, a probable error of 5% is assigned to the theory based on the behavior of other matrix elements, so that higher-order calculations are clearly necessary to get to the 1% level.

In order to investigate the behavior of MBPT for cesium in higher orders, we have recently examined the ionization energy of the valence 6s electron of cesium[15]. Because energy calculations are simpler than matrix element calculations, we were able to perform a complete calculation through third order, and also to include certain important fourth-order terms. The results are shown in Table 2.

TABLE 2 MBPT contributions to cesium 6s ionization energy (atomic units)

E_{HF}	$E^{(2)}$	$E^{(3)}$	$E^{(4)}$	E_{total}	Exp.
-.12737	-.01774	.00563	-.00370	-.14318	-.14310

As usual, the lowest order energy is rather poor, differing from experiment by 11%. There is no first order energy shift for this potential, but the second order energy gives an energy 1.4% too large in magnitude. Unfortunately, adding in the third order energy leads to a deterioration of the agreement to 2.8% too small in magnitude. This last calculation was a supercomputer scale task: even after rather severe basis set and partial wave truncations, about three hours of CRAY XMP/48 time were required. The inclusion of certain fourth-order terms that are associated with iterations of the large second-order energy shift improves the agreement to under .1%: this is probably fortuitous, as the same calculation for other valence states gives results agreeing with experiment at the .2% level. While these results are quite good, and the motivation for choosing the selected set of fourth-order energies strong, how do we know that if the complete fourth order calculation were done that the agreement with experiment would not be much poorer? Until recently, one really had to rely on one's intuition, or hand-waving arguments to address this kind of challenge. However, with supercomputing power, there is now a very direct way of finding out, which is of course to simply directly calculate all fourth-order effects. As there are well over 100 such terms for one-valence electron atoms, there is a great deal of labor involved in simply cataloging these terms, let alone doing the angular reductions and carrying out the actual coding. It is possible that some kind of automation of this phase of the calculation could be set up: certainly such automation will be necessary to go on to fifth order. It may be that, particularly if augmented by some all-orders methods, that a complete fourth or fifth order calculation of this complex system will yield results accurate at the part-per-million rather than part-per-thousand level, and that one will need to include QED effects. Increasing supercomputer power will be an essential part of any such program.

As mentioned in the introduction, the study of high-Z few-electron ions provides an example of what is only a possibility for neutral systems. In Table 3 are presented the MBPT contributions to the 3s energy of sodium-like iron through fourth-order, with fourth-order including only one of the dominant contributions discussed above. In this case, the improvement over neutral cesium is dramatic.

TABLE 3 MBPT contributions for the 3s energy of sodium-like iron(atomic units)

E_{HF}	$E^{(2)}$	$E^{(3)}$	$E^{(4)}$
-17.963529	-0.025852	0.000171	-0.000044

There is basically an additional factor 1/Z for each order of perturbation theory, so that the fourth-order energies that were playing an important role for cesium are now basically negligible in comparison with QED effects. These effects can be inferred by comparing the theory for transitions between two different states (direct ionization energies are less accessible experimentally). For the $3s$-$3p_{1/2}$ transition, the experimental splitting in atomic units (a.u.) is 1.2630 a.u.. Now, if to the results of Table 3 are added other known effects, including the Breit interaction, reduced mass, and mass polarization, the result of 1.2662 a.u. is predicted. This discrepancy clearly cannot be attributed to wave function uncertainties, but instead must be understood as a field-theoretic effect. The reason for the large size of the effect is that it is dominated by the Lamb shift, which in the one-electron case is known to behave as $Z^4\alpha^3$ a.u.. Thus a ppm effect in hydrogen becomes a 25% effect at Z=26. It is clear that there is a great deal of motivation to study the spectrum of high-Z few electron ions: they have the two desirable properties of greatly enhanced radiative corrections and extremely rapid MBPT convergence. As MBPT fits naturally into Furry representation QED, it is in principle straightforward to account for the discrepancy noted above. A discussion of this problem can be found in reference 16. The main problem is that one is at the boundary between conventional QED and the many-body problem. Radiative corrections such as those discussed in connection with the electron anomalous magnetic moment can be represented as Feynman graphs, in which each electron line represents free propagation. With such simple propagators, a great deal of experience has been built up in the evaluation of such expressions, in particular the treatment of the characteristic ultraviolet divergences of field theory. In the Furry representation, however, the propagators are much more complicated. In addition, if one wishes to make contact with MBPT expressions starting with Feynman graphs, one must carry out the integration over the fourth component of momentum associated with each closed loop using Cauchy's theorem, which corresponds to the use of time-ordered perturbation theory. Unfortunately, for the treatment of ultraviolet divergences, it is far better to not follow this procedure, and instead to treat all four components of that momentum covariantly. The complete calculation of QED effects in these ions is a very challenging task that is presently being actively investigated by a number of groups. As is in the case in so many other areas of physics, as can be seen from the other contributions to this conference, the power of supercomputers will very likely prove an integral part of this progress.

Acknowledgements: I would like to thank the organizers of this conference for the privilege of participating in such a stimulating event, and S.A. Blundell, T. Kinoshita, I. Lindgren, P.J. Mohr, D.R. Yennie, and especially W.R. Johnson for many valuable discussions. This work was supported in part by NSF grants PHY-85-03417 and PHY-86-08101, and also NSF grant PHY82-17853, supplemented by funds from NASA.

REFERENCES

1. I. Lindgren and J. Morrison, Atomic Many-Body Theory (Springer-Berlin, 1982).
2. P. B. Schwinberg, R. S. Van Dyck, and H. G. Dehmelt, Phys. Rev. Lett. 47, 1679 (1981).
3. T.Kinoshita and W. B. Lindquist, Phys. Rev. D27 853, 867, 877, 886 (1983).
4. L. S. Brown, G. Gabrielse, K. Helmerson, and J. Tan, Phys. Rev. A32, 3204 (1985).
5. G. P. Lepage, J. Comput. Phys. 27, 192 (1978).
6. Private communication from T. Kinoshita.
7. S. L. Gilbert, M. C. Noecker, P. N. Watts, and C. E. Wieman, Phys. Rev. A34, 792 (1986).
8. S. Weinberg, Phys. Rev. Lett. 19, 1264 (1967); A. Salam, in Elementary Particle Theory, edited by N. Svartholm (Almquist and Forlag, Stockholm, 1968).
9. J. P. Desclaux, Comput. Phys. Commun. 9, 31 (1975); I. P. Grant, B. J. McKenzie, P. H. Norrington, D. F. Mayers, and N. C. Pyper, Comput. Phys. Commun., 21, 207 (1980).
10. W. Muller, J. Flesch, and W. Meyer, J. Chem. Phys. 80, (1984).
11. J. Sucher, Int. J. Quantum Chem. 24, 3 (1984).
12. W.H. Furry, Phys. Rev. 81, 115 (1951).
13. W. R. Johnson, S. A. Blundell, and J. Sapirstein, Phys. Rev. A37, 307 (1988).
14. W. R. Johnson, S. A. Blundell, Z. W. Liu, and J. Sapirstein, Phys. Rev. A37, 1395 (1988).
15. S. A. Blundell, W. R. Johnson, and J. Sapirstein, submitted to Phys. Rev. A
16. W. R. Johnson, S. A. Blundell, and J. Sapirstein, Phys. Rev. A37, 2764 (1988).

SPECTROSCOPY AND STRANGENESS IN THE QUARK-GLUON PLASMA

John B. Kogut
Department of Physics, University of Illinois at Urbana-Champaign, 1110 West Green Street, Urbana, IL 61801

Abstract We discuss two simulations of QCD which require gigaflop computing. The first is the character of chiral symmetry realization in the hot quark-gluon plasma and the second is its strangeness content. Both topics are relevant to heavy-ion collisions. This work was done in collaboration with C. DeTar of Utah and D. K. Sinclair of Argonne.

I. INTRODUCTION TO THE SPECTRUM IN THE PLASMA

At high temperatures and zero baryon number density, hadronic matter is expected to undergo a phase change to a new form of matter called the quark plasma. Recently varous laboratories have mounted efforts to produce quark matter in high energy heavy ion collisions[1]. It is therefore increasingly important to develop a solid theoretical understanding of the expected structure and properties of this novel phase, not merely for its intrinsic theoretical interest, but to suggest and criticize possible experimental signals for plasma formation.

A popular model of the plasma is based upon a naive interpretation of the consequences of asymptotic freedom that obtains at extremely high temperatures. According to this "deconfinement folklore", the quark plasma is to be regarded as a gas of weakly interacting quarks and gluons. However, it is known that long-range, non-perturbative effects disrupt this simple picture even at the highest temperatures[2,3]. Knowing the long-range structure of the plasma is tantamount to knowing its large-scale composition. A great deal depends upon a phase change, plasma transport properties, multiplicities and flavors of low-energy particles, and production rates of low energy lepton pairs, to name a few.

In an effort to develop a more rigorous non-perturbative picture of the structure of the plasma, we have carried out a numerical simulation of quantum chromodynamics on small lattices [6×10^3 and 6×20] at temperatures near the phase transition. Fermions are incorporated in the staggered scheme[4] using a modification of the Illinois hybrid microcanonical algorithm[5]. We measured several hadron propagators at large space-like separation. Although our methods are well known in studies of the zero temperature hadron mass spectrum[6], to our knowledge, ours is the first deliberate application of these methods to the quark plasma. The inspiration for the present work comes from conjectures concerning the confining properties of the quark plasma and similar work on glueball analog modes of the pure gluon plasma[3,7].

With N Flavors of massless quarks, QCD is symmetric under the chiral $SU(N) \times SU(N) \times U(1) \times U(1)$ group. The $U(1) \times U(1)$ chiral symmetry is explicitly broken by the gauge anomaly, and the $SU(N) \times SU(N)$ symmetry is spontaneously broken at zero temperature, resulting in the appearance of a massless Goldstone boson. A modified version of the Goldstone theorem applies at finite temperature as well, with the Goldstone boson appearing as a zero frequency excitation at zero wave number. It has been known for some time that there must be a high temperature phase transition in QCD that restores some part of the chiral symmetry. The restoration of the $SU(N) \times SU(N)$ symmetry is signalled by a vanishing of the order parameter $<\bar{q}q>$. This effect is found in numerical simulations[8,9]. A restoration of this symmetry would allow the plasma mode vestige of the Goldstone boson, should it survive, to have a non-zero gap frequency, and would require the formation of chiral multiplets of states related by the larger symmetry. The chiral multiplet structure is determined explicitly by the valence quark assignments for the various states. Thus, for example, the plasma mode π and σ must fall in a single multiplet, the a_1 and ρ in a multiplet, and the nucleon must either be parity doubled or have zero gap frequency.

To find the plasma chiral multiplets, should they exist, requires a study of the low lying modes of excitation. They are defined by correlations of local operators A(x,t) and B(x,t):

$$S_{AB}(x,t) = \langle A(x,t)B(0,0)\rangle - \langle A(0,0)\rangle\langle B(0,0\rangle \quad (1)$$

where the average is taken over the Gibbs ensemble at temperature T. Unfortunately, lattice gauge theory has not been formulated in such a way that numerical simulations can measure finite temperature, real time response. Nonetheless, it is possible through a numerical simulation to obtain indirect information about the excitations by studying the static screening lengths, i.e. correlations between static operators at large distances and high temperatures:

$$S_{AB} = \langle \overline{A}(z)\overline{B}(0) - \langle \overline{A}(0)\rangle\langle \overline{B}(0)\rangle \quad (2)$$

where the time and transverse plane averages are given by

$$\overline{A}(z) = \lim_{L\to\infty} \int_0^\beta d\tau \int_{-L}^{L} dx \int_{-L}^{L} dy \; \frac{A(x,y,z,-i\tau)}{4L^2\beta} \quad (3)$$

The large distance behavior of this correlation,

$$S_{AB}(z) \underset{|z| \to \infty}{\simeq} b \exp[-\mu(T)|z|] \quad (4)$$

gives a screening "mass" or inverse screening length $\mu(T)$.

How is the screening phenomenon related to a real-time excitation? If we consider the dispersion relation of one of the normal modes of the plasma, given by $f(k,\omega,T) = 0$, where $k = |k|$ and ω are the real momentum and real frequency of the real-time response[1], then the screening mass obtained in[4] is found by analytic continuation to be a solution to $f[\pm i\mu(T),0,T] = 0$ for the longest mode in the channel determined by the quantum numbers of the operators A and B[3]. Just as the plasmon in an ordinary electrodynamic plasma is associated with the phenomenon of

Debye screening, so we expect low-lying excitations of the plasma to be related to the screening effects that we measure.

It is easily shown that when a symmetry is not spontaneously broken, i.e. the Gibbs ensemble is invariant under the symmetry transformation, hadron channels related by symmetry must have identical spectral properties, and, in particular, the screening masses must reveal the multiplet structure.

The first half of this report highlights our results without giving details of the numerical analysis. A complete report is published[10]. Section 2 describes our analysis briefly, and gives some of our results, and Section 3 summarizes our conclusions.

II. METHOD OF ANALYSIS AND HIGHLIGHTS OF RESULTS

Simulations were carried out at three values of the bare quark mass, namely $m = 0.05$, $m = 0.075$, and $m = 0.10$ in lattice units, at each of three values of the gauge coupling adjacent to the phase transition, namely $\beta = 5.10$, $\beta = 5.175$, and $\beta = 5.25$. In the chiral limit the first value of β is expected to lie in the low-temperature ($T < T_c$) phase, and the last two, in the high-temperature ($T > T_c$) phase. The bulk of the measurements were carried out on the smaller lattice, namely 6×10^3.

Screening masses were measured for a variety of color singlet channels with the quantum numbers of the π, σ, ρ, b_1 and a_1 mesons and both parity states of the lowest lying baryon. These masses were measured by calculating quark propagators on lattices obtained by doubling the gauge field configuration in a spatial direction.

A careful analysis is required to determine the systematic and statistical errors in the screening masses. Systematic errors arise from finite size effects, a neglect of higher spectral components, and a variety of correlations[10].

Shown in Fig. 1 is a sample plot of the static correlation[2] for the π meson channel in the high temperature phase. The fit is to the form

$$S_\pi(z) = b\,\exp(-\mu z) + b\,\exp[-\mu(L-z)] \tag{5}$$

where L is the length of the lattice in the z direction. In this case the fit begins at $z = 4$ and runs to the largest value of z, so as to minimize the effect of higher spectral components. The fit to the data running to $z = 20$ gives a screening mass of 0.765 with a χ^2 of 29 for 15 df. The fit to the data from the 6×10^3 lattice alone with a far larger data sample gives a screening mass of 0.754 ± 0.003 with $\chi^2 = 20$ for 5 df in lattice units. (In this and other channels, only values from the smaller lattice are used in the extrapolation to the chiral limit.)

Thus the static correlation shows a remarkably clean fit to a single spectral component over six decades. If there were a $q\bar{q}$ component in this channel it would have a continuum threshold screening mass of at least twice the lowest fermion Matsubara frequency, namely $2\pi T$. In the present lattice units $T = 1/6$, so this threshold corresponds to a screening mass higher than 1. Since the pionic spectral component seen in Fig. 1 is obviously of a considerably lower screening mass, it cannot be due to such a quark continuum. However, we cannot rule out the possibility that such a continuum occurs in addition to the pionic mode as a higher spectral component.

Other channels were analyzed in a similar way. The fitting functions for the other channels include oscillating terms according to the requirements of the staggered fermion scheme[11]. Fig. 2 plots the results for the π and σ mesons at $b = 5.25$ as a function of bare quark mass. The data is consistent with the expected degeneracy in the chiral limit. The data is also consistent with the conclusion that the pionic mode is not a Goldstone boson for $T > T_c$.

Results for other hadronic channels at the three values of β were similarly extrapolated to zero quark mass. Where statistically allowed, the intercepts of the expected multiplets were equated. Results for extrapolations to the chiral limit of all of the channels are collected in Fig. 3. The shaded region in this figure indicates the possible location of the phase transition in the chiral limit[8,12]. It is evident that the ρ and a_1 fall into the expected multiplets above the phase transition and that the lowest lying baryonic mode chooses parity doubling, instead of a zero mass. We also find some indication in the $\beta = 5.25$ data that the two helicity states of the vector/axial vector meson are split.

III. SYMMETRY REALIZATION IN THE PLASMA

Our results are consistent with a restoration of an $SU(N) \times SU(N)$ chiral symmetry. In particular, we find good numerical evidence in the chiral limit of our model,

1. that there is a cleanly identifiable pion plasma mode in the high temperature phase;
2. that the expected π-σ and ρ-a_1 multiplets occur in the plasma;
3. and finally, that there are parity doubled baryon plasma modes with finite screening length for $T > T_c$.

A partial restoration of the U(1) axial symmetry, associated with the diminution of the gauge anomaly, would require a further parity doubling of the π -σ multiplet[13]. However, we have not examined the relevant channels, and so cannot draw any conclusions regarding the fate of the U(1) symmetry.

Our results strongly suggest the existence of hadronic modes in the plasma screening spectrum. Whether these modes are also important as real-time excitations of the plasma is an urgent question that will probably not be answered soon by lattice gauge theory. Nonetheless, their appearance in the screening spectrum deals a serious blow to the naive deconfinement picture and requires a reconsideration of several of its predictions of the experimental plasma signature.

IV. THERMODYNAMICS AND A REALISTIC QUARK SPECTRUM

Now we present data for a relatively realistic quark spectrum, an isodoublet of light quarks and a heavier strange quark, and we display metastable states at the hadronic matter/plasma transition. We also measure the gluonic, isodoublet and strange quark contributions to the system's internal energy and find discontinuous jumps in each internal energy which are in the ratio 8:4:1. This suggests that the strangeness content in a quark-gluon plasma made in a heavy-ion collision should be considerable and could provide a signal for the creation of the plasma itself.

Before discussing results, consider some issues of lattice technology and algorithms. We have simulated small lattices, 4×8^3, where the temperature T is related to the lattice spacing "a" and the temporal extent $N_t = 4$ by $aT = 1/N_t$. Quark masses were chosen to be $am_{u,d} = .0125$ for the isodoublet and $am_s = 0.25$ for the strange quark. If $T_c \simeq 155$ MeV as favored by phenomenology[14] as well as crude lattice spectrum calculations,[15] then the bare mass of the isodoublet is 7.75 MeV and the bare mass of the strange quark is 155 MeV. These masses are near the quark mass estimates based on current algebra.[16] Therefore, aside from serious questions of assessing finite lattice spacing and finite volume effects, these simulations are quite realistic.

Our numerical results were obtained using a version of the hybrid algorithm which propagates the gluon fields in computer time t while accounting for the effects of quarks through additional stochastic terms in the gluon equation of motion.[17] The gluon equation of motion is discretized with basic time steps $dt = 0.01$ and systematic errors of $0(dt)^2$ have been observed in the algorithm's time averages. To accomodate an isodoublet of quarks a random complex field $\eta_{u,d}$ is placed on the sites of the lattice and an independent complex field η_s is also placed on the lattice to simulate the effects of a strange quark. Two terms, each bilinear in $\eta_{u,d}$ and η_s, then appear in the gluon equation of motion and the strength of each term can be adjusted continuously so that an isodoublet and a strange quark are simulated.

V. ORDER OF THE TRANSITION AND STRANGENESS CONTENT

Now consider the results of our simulations. Our first task was to find the chiral symmetry restoring transition which separates the hadronic matter from the quark-gluon plasma phase. Our previous simulations with a light isodoublet of quarks placed the critical coupling at $\beta_c = 6/g_c^2 \simeq 5.27\text{-}5.29$ where metastable states were displayed.[18] We expected to find a first order transition in the theory with a realistic quark spectrum at a β_c shifted to a smaller value due to the additional color screening caused by the strange quark. This effect was observed. In Fig. 4 we show the time evolution of the observables $\langle\bar{\psi}\psi\rangle$, the chiral order parameter of the light isodoublet, WL, the Wilson line and S_O, the average plaquette of Wilson's lattice gauge action, at $\beta = 5.25$. The two curves correspond to a hot and cold start and we see both configurations evolve into a hot, quark-plasma state where $\langle\bar{\psi}\psi\rangle$ is almost zero and WL is relatively large. The total number of sweeps in Fig. 4 is 10,000 and the observables were averaged into bins 100 sweeps wide. The time evolution of hot and cold starts at $\beta = 5.20$ are shown in Fig. 5 for 30,000 sweeps and a clear two state signal was found. This is our best evidence for the first order character of the transition and metastable states. It is hard to judge the degree of metastability observed here except to note that for β chosen away from β_c the time correlation length in the algorithm (the number of sweeps required to generate a statistically independent configuration) is typically 100-300 sweeps, so a 30,000 sweep run would generate from 100 to 300 new gauge field configurations. Finally, in Fig. 6 we show a metastability search at $\beta = 5.15$ and find that the states evolve into a unique state of ordinary hadronic matter where $\langle\bar{\psi}\psi\rangle$ is large and WL is small. Only 10,000 sweeps were needed in Fig. 6.

In summary, SU(3) gauge theory with a realistic quark spectrum has a first order chiral transition on a 4×8^3 lattice at a coupling $\beta = 5.20\pm0.05$.

To understand the character of this transition we also measured other observables such as the pressure and energy densities. The internal energy measurements are shown in Fig. 7. We note from the figure that the three internal energies also jump at $\beta = 5.20$ as expected of a strong first order transition. Unfortunately the extraction of physically interesting numbers from Fig. 7 requires some analysis and assumptions, due to the presence of large finite size effects on small lattices. First we note that the dimensionless energy densities ε/T^4 are substantial fractions of the Stephan-Boltzmann free field internal energies on lattices of this size.[19] For the gluon internal energy, the ratio of the internal energy shown in Fig. 7 at β slightly above 5.20 to the Stephan-Boltzmann value for a free lattice bose gas of 16 degrees of freedom (8 for color, times 2 for spin) is 2.20±0.20. For the isodoublet the ratio is 0.85±0.15 and for the heavier strange quark it is 0.41±.0.06. Presumably this ratio is smaller for the heavy quark because of its considerable mass, $m_s \simeq T_c$. Next we note that the free bose gas on a 4×8^3 lattice has an internal energy which exceeds the space-time continuum Stephan-Boltzmann value ($\varepsilon_b/T^4 = \pi^2/30$) by a factor of 1.40 due to finite size effects and a free quark gas exceeds its Stephan-Boltzmann value ($\varepsilon_f/T^4 = 7\pi^2/60$) by a factor of 1.86 for am = 0.0125 and by a factor of 1.69 for am = 0.25.[14] If we correct for these finite size effects by dividing our data by these numerical factors we have,

$$\begin{aligned} \Delta\varepsilon_g &= (12.00\pm2.00)\ T^4 \\ \Delta\varepsilon_{u,d} &= (6.30\pm1.00)\ T^4 \\ \Delta\varepsilon_s &= (1.70\pm0.25)\ T^4 \end{aligned} \qquad (6)$$

The major interest in Eq. (1) is that each internal energy is large and they stand in the ratio 8:4:1 (approximately). Therefore, this lattice gauge theory simuation predicts considerable $s\bar{s}$ production in heavy ion collisions.

Models of the hadronization of the quark-gluon plasma suggest that a relatively large fraction of $s\bar{s}$ quarks should lead to considerable anti-hyperon production.[20] Model builders and experimentalists should find Eq. (6) provocative and useful in quark hadronization studies of the evolution of the system's final state.

REFERENCES

1. For a recent conference proceedings, see *Proceedings of the Fifth International Conference on Ultra-Relativistic Nucleus-Nucleus Collisions: Asilomar Conference Center, April 13-17, 1986,* ed. L. S. Schroeder and M. Gyulassy, Nucl. Phys. **A461** Nos. 1,2 (1987). For a review of QCD calculations, see H. Satz. Ann. Rev. Nucl. Part. Sci. **35**, 245 (1985).
2. D. J. Gross, R. D. Pisarski, and L. G. Yaffe, Rev. Mod. Phys. **53**, 43 (1981); K. Kajantie and J. Kapusta, Phys. Lett. **110B**, 299 (1982); T. Applequist and R. Pisarski, Phys. Rev. **D23**, 2305 (1981); S. Nadkharni, Phys. Rev. **D27**, 917 (1983); J. Polónyi and H. Wyld, Urbana report ILL-(TH)-85-23 (unpublished) and MIT report CTP #1458 (1987, unpublished); U. Heinz, K. Kajantie, and T. Toimela, BNL Report (1986, unpublished); T. H. Hansson and I. Zahed, Stony Brook preprint (1987, unpublished).
3. C. E. DeTar, Phys. Rev. **D32**, 276 (1985).
4. J. Kogut and L. Susskind, Phys. Rev. **D11**, 395 (1975).
5. J. Polónyi and H. W. Wyld, Phys. Rev. Lett. **51**, 2257 (1983); S. Duane and J. B. Kogut, *ibid.* **55**, 2774 (1985); J. B. Kogut. Nucl. Phys. **B270** FS16, 169 (1986).
6. H. Hamber and G. Parisi, Phys. Rev. **D27**, 208 (1983); J. P. Gilchrist, H. Schneider, G. Scheirholz, and M. Teper, Phys. Lett. **136B**, 87 (1984); M. Fukugita, S. Ohta, Y. Oyanagi, and A. Ukawa, KEK report KEK-TH 150 (1987).
7. T. A. DeGrant and C. E. DeTar, Phys. Rev. **D34**, (1986).

8. J. B. Kogut and D. K. Sinclair, Illinois report ILL-(TH)-86-46 (1986).
9. M. Fukugita and A. Ukawa, Phys. Rev. Lett. **57**, 503 (1986): R. Gupta et al. *ibid.* 2621 (1986); S. A. Gottleib et al., UCSD Report (1987).
10. C. DeTar and J. Kogut, Phys. Rev. **D36**, 2828 (1987).
11. H. kluberg-Stern, A. Morel, O. Napoly, and B. Petersson, Nucl. Phys. **B220**, 447 (1983); J. Kogut et al., Nucl. Phys. **B225** [FS9], 326 (1983).
12. E. V. E. Kovacs, D. K. Sinclair, and J. B. Kogut, Phys. Rev. Letters **58**, 751 (1987).
13. R. Pisarski and F. Wilczek, Phys. Rev. **D29**, 338 (1984).
14. J. Cleymans et. al., Phys. Rev. **130**, 217 (1986).
15. S. Gottlieb, W. Liu, D. Toussaint, R. L. Renken and R. L. Sugar, Phys. Rev. Lett. **59**, 1513 (1987). M. Grady, D. K. Sinclair and J. B. Kogut, Physics Letters 200B, 149 (1988).
16. S. Weinberg, in *Festschift for I. Rabi*, edited by L. Motz (N.Y. Academy of Sciences, N.Y., 1977). p. 243.
17. S. Gottlieb, W. Liu, D. Toussaint, R. L. Renken and R. L. Sugar, Phys. Rev. **D35**, 2531 (1987).
18. J. B. Kogut and D. K. Sinclair, University of Illinois, ILL-(TH) -87-46 (to appear in Nucl. Phys.)
19. J. Engels, F. Karsch and H. Satz, Nucl. Phys. **B205** [FS5], 239 (1982). The factors 1.86 and 1.69 used to obtain Eq. (1) from Fig. 4 are for staggered fermions (private communication, F. Karsch, H. W. Wyld.)
20. P. Koch, B. Mueller and J. Rafelski, Phys. Rep. **142**, 167 (1986).

FIGURE CAPTIONS

1. The static correlation[2] in the pion channel, averaged over x and y, as function of the separation z in the high temperature phase. The points are from the numerical simulation at $\beta = 5.25$ and quark mass $m = 0.05$ in lattice units. The errors are smaller than the symbol sizes.
2. Screening masses for the π and σ modes in the high temperature phase ($\beta = 5.25$) as a function of the bare quark mass. All masses are in lattice units. The curves show a linear extrapolation of the π and quadratic extrapolation of the σ screening masses to a common intercept in the chiral limit.
3. Screening masses, as extrapolated to the chiral limit, for the π, σ, ρ and a_1 meson plasma modes, and the lowest even parity (N_+) and odd parity (N_-) baryon plasma modes, plotted as a function of the gauge coupling β, expressed in units of the temperature. Increasing β corresponds to increasing temperature. The shaded region indicates the possible location of the phase transition. The notation "0" and "1" beside the highest temperature vector/axial vector meson mass indicates the helicity assignment.
4. Time evolution of $\langle\bar{\psi}\psi\rangle$ the Wilson line WL and the average plaquette S_0 from hot and cold starts at $\beta = 5.25$. 10,000 sweeps.
5. Same as 1 except $\beta = 5.20$, and 30,000 sweeps.
6. Same as 1 except $\beta = 5.15$.
7. Internal energies, ε/T^4, for gluons (∇) light quarks (o) and heavy quark (·).

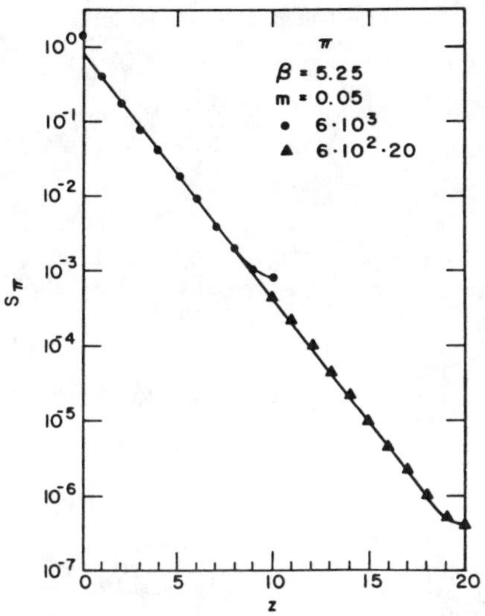

Fig. 1

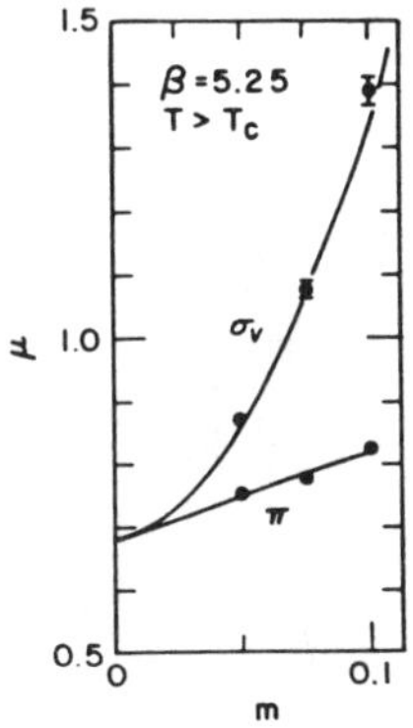

Fig. 2

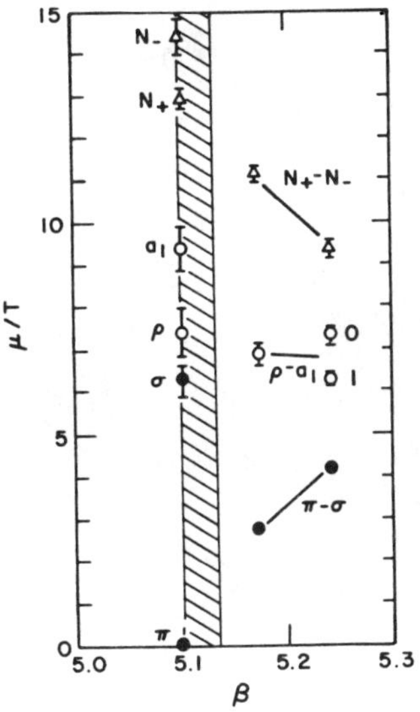

Fig. 3

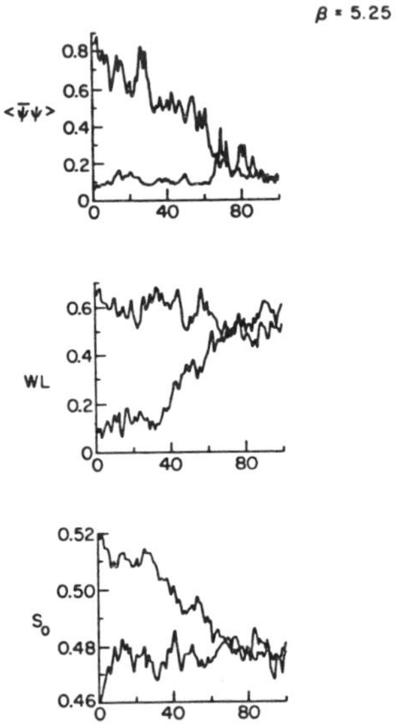

Fig. 4

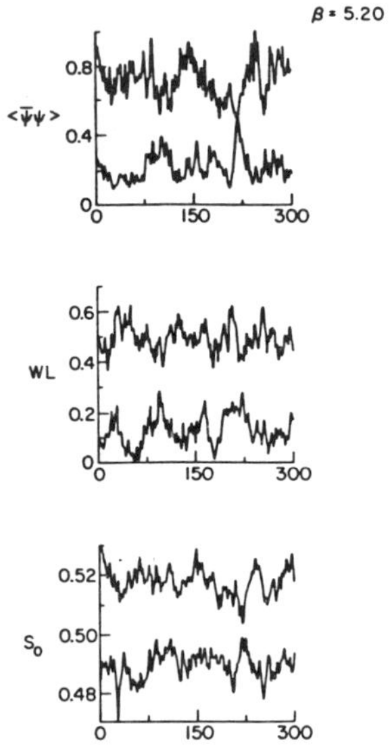

Fig. 5

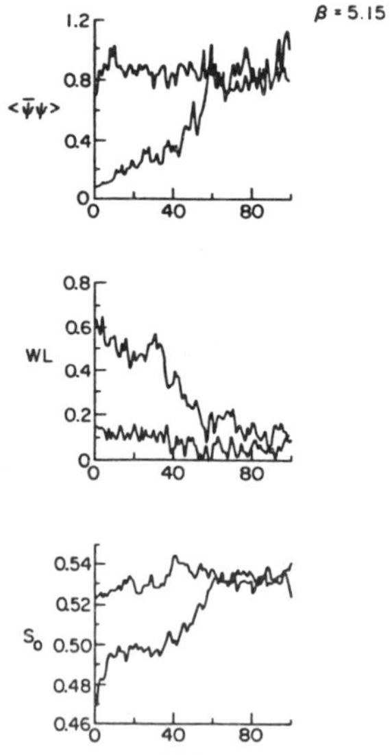

Fig. 6

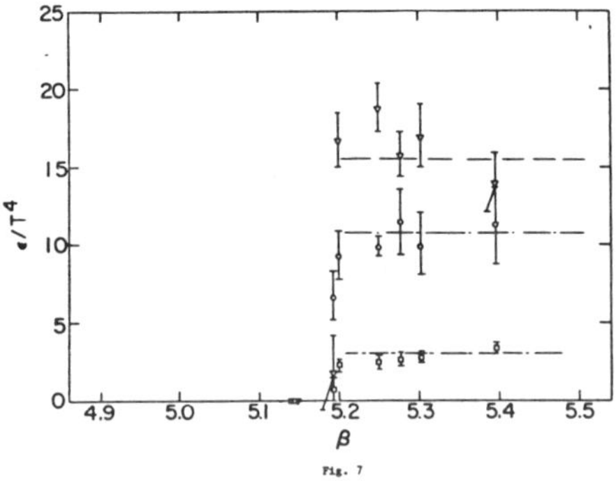

Fig. 7

GREEN'S FUNCTION MONTE CARLO WITH "SHADOW" TRIAL FUNCTIONS

S. A. Vitiello, K. J. Runge, and M. H. Kalos
Courant Institute of Mathematical Sciences,
New York University, New York, NY 10012, USA

Abstract A new class of trial wave functions for the investigation of quantum crystals and liquids, the "shadow" wave functions, is presented. These trial functions are symmetric under particle interchange and translationally invariant. Some of our results obtained in variational calculations are discussed. We show how "shadow" wave functions may be used for importance sampling in Green's function Monte Carlo: one alternates Metropolis moves for "shadow" coordinates with GFMC moves for the real coordinates.

I. INTRODUCTION

Exploration of the properties of extensive quantum systems is very much helped by the insight that comes from analytic trial functions. There is a variety of problems, such as the properties of quantum crystals or quantum vortices whose elucidation requires some explicit many-body ansatz. The problems associated with quantum crystals are long-standing and well known: the usual trial functions suitable for variational calculations and for importance sampling in a Green's function Monte Carlo (GFMC)[1] are of the Nosanow-Jastrow[2] type, in which the structure of the crystal is built in *a priori* by localizing the particles to a lattice through a one-body pseudo-potential. Although this gives a reasonable variational energy for a crystal, and accurate and robust energies from GFMC calculations, it remains unsatisfactory. For one thing, it leaves open the question of whether certain subtle properties of the crystal, such as the existence and magnitude of a condensate, can be computed correctly. For another, the kind of qualitative understanding that can come only from the interplay of numerical results and more traditional theory is also questionable when the trial functions fail to satisfy fundamental properties of the system such as Bose symmetry and translational invariance.

We have discovered a new class of trial wave functions,[3] to which we have given the name "shadow" wave functions, that remove these difficulties for quantum crystals. We conjecture that they will also prove to be the basis for the construction of trial wave functions for quantum vortices. The qualitative idea is simple: the wave function is a product of a convolution of a very simple many-body function (that now describes the correlations of fictitious or "shadow" particles) by another such simple function. The rationale for introducing such a structure can be given in several ways. A purely phenomenological justification comes from the idea that one must allow the two-body correlation for the crystal to be determined primarily from the pair forces, so that the crystalline order now comes from the (possibly stronger) pair-wise correlation of the shadow particles and the coupling of real to shadow particles. Another way of looking at it is to adopt a path-integral point of view.[4] Then the center of mass of a path followed by one particle can be seen as a quasi-classical (or "shadow") particle that repels other such particles with a large core. Finally, the formalism of Green's function Monte Carlo suggests the idea directly of multiplying simple trial functions with a kernel that approximates to the inverse of the Hamiltonian and integrating.

In section II we give the explicit definition of the "shadow" wave function. Section III discusses the system under consideration, the form of the "shadow" wave function applied to variational calculations and some of our results. In section IV we show that Green's function Monte Carlo together with Metropolis algorithm converges to the exact ground state if a "shadow" wave function is used in importance sampling.

II. THE "SHADOW" WAVE FUNCTION

The variational "shadow" wave function $\Psi_T(R)$ for a system with N particles, $R \equiv \{\vec{r}_1, \vec{r}_2, \ldots, \vec{r}_N\}$, is constructed by considering a set of auxiliary or shadow variables $S \equiv \{\vec{s}_1, \vec{s}_2, \ldots, \vec{s}_N\}$ through

$$\Psi_T(R) = \int \Xi(R,S)dS \ . \tag{1}$$

The function $\Xi(R,S)$ used here is the product of three terms

$$\Xi(R,S) = \psi_r(R) \quad \prod_k \theta(\vec{r}_k - \vec{s}_k) \quad \psi_s(S) \ . \tag{2}$$

The factors ψ are of the Jastrow form; thus, for the real particles it is given by

$$\psi_r(R) = \prod_{i<j} e^{-\frac{1}{2}u(r_{ij})} \ , \tag{3}$$

where $r_{ij} = |\vec{r}_i - \vec{r}_j|$ is the distance between particles i and j and the pseudo-potential $u(r_{ij})$ has for example the McMillan[5] form,

$$u(r) = \left(\frac{b}{r}\right)^5 \ , \tag{4}$$

with b taken as a variational parameter. The second factor in Eq. (2), involving $\theta(\vec{r}_k - \vec{s}_k)$ provides a direct one-to-one coupling between a particle and shadow variables. In this work it is chosen as a Gaussian:

$$\theta(\vec{r}_k - \vec{s}_k) = e^{-C \ (\vec{r}_k - \vec{s}_k)^2} \ , \tag{5}$$

where C is a variational parameter. Finally, $\psi_s(S)$ is a Jastrow factor that couples the shadow variables and is written as

$$\psi_s(S) = \prod_{i<j} e^{-v(s_{ij})} \ , \tag{6}$$

where $s_{ij} = |\vec{s}_i - \vec{s}_j|$. The model potential $v(s_{ij})$ selected for the shadows is

$$v(s) = \left(\frac{b_{sh}}{s}\right)^n , \tag{7}$$

and has two variational parameters b_{sh} and n.

III. VARIATIONAL CALCULATIONS

The Hamiltonian of N ^{4}He atoms with mass m is taken as

$$H = -\frac{\hbar^2}{2m}\sum_{i=1}^{N}\nabla_i^2 + \sum_{i<j} V(r_{ij}) . \tag{8}$$

We use the accurate two-body potential $V(r_{ij})$ proposed by Aziz *et al.*[6] This potential has been extensively used and has yielded excellent results for condensed helium.[7]

The ground state energy has an upper bound given by

$$E_0 \leq E_T = \frac{\int \Psi_T H \Psi_T dR}{\int |\Psi_T|^2 dR} . \tag{9}$$

This equation can be rewritten in terms of equations (1) and (2) so that it reads:

$$E_T = \iiint \left\{ \frac{H\psi_r(R)\prod_k \theta(\vec{r}_k - \vec{s}\,'_k)}{\psi_r(R)\prod_k \theta(\vec{r}_k - \vec{s}\,'_k)} \right\} p(R,S,S')dRdSdS' , \tag{10}$$

where $p(R,S,S')$ is given by

$$p(R,S,S') = \frac{\Xi(R,S)\Xi(R,S')}{\iiint \Xi(R,S)\Xi(R,S')dRdSdS'} . \tag{11}$$

The integral in Eq.(10) can be evaluated by the Metropolis Monte Carlo algorithm,[8] with a straightforward modification to take into account the shadow coordinates. The probability density function p of Eq.(11) depends on the configuration R of the real particles and on the configurations S and S' of the shadow particles. To sample $p(R,S,S')$ it will be convenient to subdivide each Metropolis step in two parts. In one of them, new configurations for the shadow coordinates are generated conditional on R. In the other one, a new configuration, R, is sampled while keeping S and S' fixed. A more formal description is as follows.

Let $T_s(R|R', S, S')$ be a stochastic kernel for R, conditional on (R', S, S') and define $A_s(R|R', S, S')$, for any given function $f(R, S, S') \geq 0$, as

$$A_s(R|R', S, S') = min(1, \frac{T_s(R'|R, S, S')f(R, S, S')}{T_s(R|R', S, S')f(R', S, S')}) \,. \tag{12}$$

Then

$$K_s(R|R', S, S') = A_s(R|R', S, S')\, T_s(R|R', S, S') + \delta(R - R') \times \\ \times \int [1 - A_s(R|R', S, S')] T_s(R|R', S, S') dR, \tag{13}$$

is also a stochastic kernel such that for every S and S' we have

$$\mathcal{K}_s[f(R, S, S')] = \int K_s(R|R', S, S') f(R', S, S')\, dR' \\ = f(R, S, S') \,. \tag{14}$$

For the generation of new configurations of shadow coordinates we consider the stochastic transition density $T_r(S|S', R)$ for a trial move to S from S' conditional on R. The acceptance probability for this move is given by

$$A_r(S|S', R) = min(1, \frac{T_r(S'|S, R)\Xi(R, S)}{T_r(S|S', R)\Xi(R, S')}) \,. \tag{15}$$

So the probability density function $K_r(S|S', R)$ can be written as

$$K_r(S|S', R) = A_r(S|S', R)\, T_r(S|S', R) \\ + \delta(S - S') \int [1 - A_r(S|S', R)]\, T_r(S|S', R) dS. \tag{16}$$

For every R this kernel, which is also stochastic, gives

$$\mathcal{K}_r[\Xi(R,S)] = \int K_r(S|S',R)\Xi(R,S')\, dS' = \Xi(R,S) \; . \qquad (17)$$

Finally for $f(R,S,S') = \Xi(R,S)\Xi(R,S')$ we can write

$$\mathcal{K}_s^l\, [\, (\mathcal{K}_r^m \Xi(R,S))\; (\mathcal{K}_r^n \Xi(R,S'))\,] = \Xi(R,S)\Xi(R,S') \; . \qquad (18)$$

Although we get excellent results for l, m, and n equal to 1, we have the possibility to choose other values to minimize the variance of the results.

As usual the transition densities are chosen to be constants so that in each trial move the new coordinates are sampled uniformly in a domain which is a cube centered in the old ones. T_s and T_r are chosen so that the acceptance is nearly 50%.

Finally, the variational energy, the $9N$ dimensional integral of Eq.(10), can be computed by taking the average of the estimator $\left\{ \frac{H\psi_r(R)\prod_k \theta(\vec{r}_k - \vec{s}\,'_k)}{\psi_r(R)\prod_k \theta(\vec{r}_k - \vec{s}\,'_k)} \right\}$ with respect to the probability density function $p(R,S,S')$. This average could also be computed using S instead of S', so that we may use a linear combination of both to reduce the the variance.

All the computations started from a fcc lattice (for both the real and the shadow particles.) For appropriate choices of the variational parameters, b, b_{sh}, and C, the system "melts" from the initial configuration into a fluid, while for other choices the system remains a solid.

In the fluid phase the variational energy per atom was computed at the equilibrium density $\rho\sigma^3 = 0.365$ ($\sigma = 2.556$Å). Table I contains the results. For comparison purposes in the first line of Table II we give the energy for the equilibrium density that is computed using a pure Jastrow wave function. A liquid state is very quickly obtained (for both real and shadow particles). The pair correlation functions $g(r)$ at the equilibrium density for both real and shadow particles are plotted in Fig. 1.

Table I. Variational energies in K per atom obtained with the shadow wave function for the liquid and solid phases at several densities. The b parameters are in units of σ while the one of C is σ^{-2}.

$\rho\sigma^3$	E_T	Parameters			
		b	b_{sh}	n	C
		liquid			
0.365	-6.061 ±0.025	1.13	1.40	5	4.0
		crystal			
0.491	-5.004 ±0.055	1.10	1.70	5	4.8
0.55	-3.540 ±0.028	1.10	1.68	5	5.7
0.55	-3.549 ±0.050	1.10	1.35	9	5.7
0.55	-3.563 ±0.031	1.09	1.28	12	5.7

Table II. Results for liquid and solid ^{4}He determined using a pure Jastrow (liquid) and a Nosanow-Jastrow (crystal) wave functions. Units are the same of those of Table I.

$\rho\sigma^3$	E_T	Lindemann ratio	Parameters	
			b	C
		liquid		
0.365	-5.717 ±0.021	–	1.20	–
		crystal		
0.55	-3.361 ±0.019	0.22	1.10	4.8

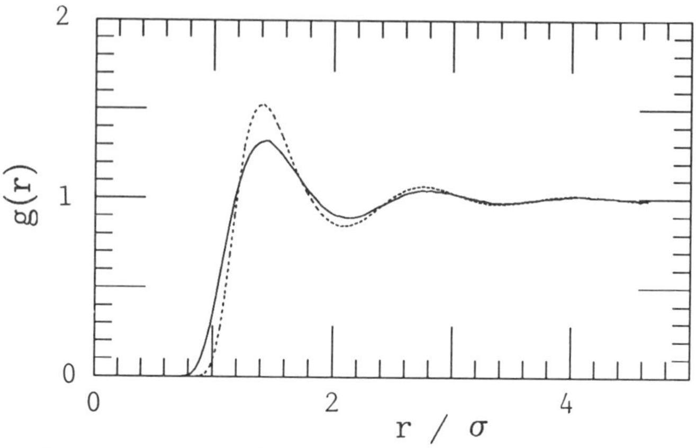

FIGURE 1. Radial distribution functions $g(r)$ computed at $\rho\sigma^3 = 0.365$. The solid line gives the result for the real particles. The pair correlation for the shadows is shown by the dotted line.

For the crystal phase of ^{4}He the variational energy is computed at the densities $\rho\sigma^3 = 0.55$ and 0.491. The latter is the melting density determined by GFMC.[7] At $\rho\sigma^3 = 0.55$ three different power laws have been tested for the correlation $v(s)$, Eq.(7). The results are shown in Table I; they are insensitive to the exact form of the shadow interaction. The second line of Table II presents the value of the energy that we have also computed using the standard Nosanow-Jastrow wave function, that is a Jastrow factor multiplied by a product of Gaussian-like localization factors that bind the atoms to fcc lattice sites. Note that the shadow variational results are significantly lower.

In the crystal phase we have computed Lindmann's ratio. This quantity is defined as the quotient of the square root of the second moment of the single particle distribution function divided by the nearest-neighbor distance. Results are shown in Table II and III, with Table II containing the values obtained from a Nosanow-Jastrow wave function. Lindmann ratios of the shadow particles are given in Table III.

Table III. Lindemann ratio obtained with the shadow wave function at two densities.

$\rho\sigma^3$	n	Lindemann ratio real	Lindemann ratio shadow
0.491	5	0.25	0.18
0.55	5	0.23	0.16
0.55	9	0.23	0.13
0.55	12	0.22	0.13

We stress that the same functional form of the wave function is used in both liquid and solid phases. The phase and measurable properties of the system are determined by the choice of the variational parameters in the wave function.

IV. GREEN'S FUNCTION MONTE CARLO WITH "SHADOW" FUNCTIONS

In the Green's function Monte Carlo method the integral form of Schrödinger's equation is solved for the ground state by iteration. The importance sampling transformation requires sampling the Green's function $G(R, R')$ of the system multiplied by $\frac{\Psi_T(R)}{\Psi_T(R')}$, so that a sequence of configurations $\{R^{(n)}\}$ can be drawn from $\Psi_T\Psi^{(n)}$. In the limit of large n, $\Psi^{(n)}$ converges to Ψ_0, the lowest state not orthogonal to Ψ_T.

With a set of configurations $\{R\}$ drawn from $\Psi_T\Psi_0$ the energy can be easily computed. From Schrödinger's equation we have

$$E = \frac{\int \Psi_T(R) H \Psi_0(R)\, dR}{\int \Psi_T(R)\Psi_0(R)\, dR} \,. \tag{19}$$

which can be rewritten in a more convenient way for our computations as

$$E = \frac{\int \left[\frac{H\Psi_T(R)}{\Psi_T(R)}\right] \Psi_T(R)\Psi_0(R)\, dR}{\int \Psi_T(R)\Psi_0(R)\, dR} \,. \tag{20}$$

A different approach is needed when we seek to use a "shadow" wave function as an importance function. It would be impractical to carry out the integration required by Eq.(1) for every value of R. Fortunately, since Monte Carlo methods naturally integrate over auxiliary variables, it is not necessary. Instead we arrange to sample a joint density in R and S that has the asymptotic value

$$\Xi(R,S)\Psi_0(R)\ . \tag{21}$$

After convergence has been reached, the set of M configurations $\{R_m, S_m\}$, can be used to compute the energy from the expression

$$E = \frac{1}{M} < \sum_{m=1}^{M} \frac{H\Xi(R_m,S_m)}{\Xi(R_m,S_m)} > \ . \tag{22}$$

It is natural and efficient to start the calculation with a converged population of configurations $\{R_m, S_m\}$ obtained in a variational calculation. Of course, $\{R_m, S'_m\}$ could be used as well. Once we have one configuration (R_m, S_m), one or more generations can be carried over taking $\Xi(R, S_m)$ as the importance function for this fixed S_m. After a given number q of generations, it is necessary to iterate the S values by the Metropolis algorithm. The reason for doing this is that as the R distribution tends from $\Xi(R,S)\Psi_T(R)$ toward $\Xi(R,S)\Psi_0(R)$, the S values must be resampled so as to be drawn correctly from the final joint distribution in R and S. In the following we show explicitly that by alternating Green's function Monte Carlo to produce populations of configurations for the real particles $\{R\}$, and Metropolis algorithm to draw new configurations for the shadow coordinates, one gets the required convergence.

First, let us define an integral operator $\mathcal{G}_S$ which acts on a function $\ell(R,S)$ in the following way

$$\mathcal{G}_s\ell(R,S) = \Xi(R,S)\int \frac{G(R,R')}{\Xi(R',S)}\,\ell(R',S)\,dR'\ , \tag{23}$$

where the kernel $G(R,R')$ is the inverse H^{-1} of the Hamiltonian operator that could be constructed using the complete set of eigenfunctions Ψ_k of H:

$$G(R,R') = \sum_k \frac{\Psi_k(R)\Psi_k(R')}{E_k} \; . \tag{24}$$

In particular, the result of applying the operator $\mathcal{G}_s$ to the product $\Xi(R,S)\phi(R)$ where $\phi(R)$ can be expanded as $\sum_k c_k \Psi_k(R)$ is given by:

$$\mathcal{G}_s[\Xi(R,S)\phi(R)] = \Xi(R,S)\sum_k \frac{c_k}{E_k}\Psi_k(R) \; . \tag{25}$$

Now we can demonstrate that the ground-state wave function will be obtained by successive application of $\mathcal{G}_s$ and the operator $\mathcal{K}_r$, Eq.(17),

$$\begin{aligned}(\mathcal{K}_r^p \mathcal{G}_s^q)^n \Xi(R,S)\phi(R) &= \Xi(R,S)\sum_k \frac{c_k}{E_k^{qn}}\Psi_k(R) \\ &\underset{n\to\infty}{\longrightarrow} \Xi(R,S)\frac{c_0}{E_0^{qn}}\Psi_0(R) \; , \end{aligned} \tag{26}$$

where q is the number of generations of GFMC for fixed S before p Metropolis interactions on the shadows with fixed R. We stress that no R moves are performed using Metropolis method. Again, reasonable values would be $p = q = 1$. However the best values for q and p are those that give the lowest variance. Computations where "shadow" wave functions are used as importance functions are now under way.

Although we have obtained very encouraging results so far for the variational energies of fluid and crystal states, we await the results of Green's function Monte Carlo for more definitive answers in both three and two dimensions. We are also exploring the theoretical questions that relate to the use of "shadow" wave functions to vortices and other excited states. It is clear that these are very demanding computations in terms of Monte Carlo techniques and of computer capacity. Gigaflops computers are a necessary tool for serious experimentation, not to say accurate results of interest to physics.

Acknowledgements: This work was partially supported by the Condensed Matter Theory Program of the National Science Foundation under Grant DMR-8419083, by the Advanced Scientific Computing Program of the NSF under Grant ASC-8715641 and by Conselho Nacional de Desenvolvimento Científico e Tecnológico, Brazil. This research was conducted using the Cornell National Supercomputer Facility, a resource of the Center for Theory and Simulation In Science and Engineering at Cornell University, which is funded in part by the National Science Foundation, New York State, and the IBM Corporation.

REFERENCES

1. P. A. Whitlock, D. M. Ceperley, G. V. Chester and M. H. Kalos, *Phys. Rev. B* **19**, 5598 (1979).
2. L. H. Nosanow, *Phys. Rev. Lett.* **13**, 270 (1964).
3. Silvio Vitiello, Karl Runge and Malvin H. Kalos, *Phys. Rev. Lett.* **60**, 1970 (1988).
4. R. P. Feynman, *Statistical Mechanics*, W. A. Benjamin, Inc., Reading, 1972.
5. W. L. McMillan, *Phys. Rev.* **138**, A442 (1965).
6. R. A. Aziz, V. P. S. Nain, J. S. Carley, W. L. Taylor and G. T. McConville, *J. Chem. Phys.* **70**, 4330 (1979).
7. M. H. Kalos, M. A. Lee, P. A. Whitlock and G. V. Chester, *Phys. Rev. B* **24**, 115 (1981).
8. N. Metropolis, A. W. Rosenbluth, M. N. Rosenbluth, A. H. Teller and E. Teller, *J. Chem. Phys.* **21**, 1087 (1953).

DRILLING HOLES THROUGH NUCLEI, HELIUM DROPS, AND FERMI GASES*

STEVEN C. PIEPER
Argonne National Laboratory, Argonne, IL 60439-4843 USA

Abstract The probability of finding a specified number of nucleons in a cylinder passing through a nucleus is calculated by Monte Carlo techniques. The resulting distribution is compared to several analytical forms. The distributions are used to compute the distribution of transverse energy production in ultra-relativistic proton-nucleus collisions.

INTRODUCTION

Experiments at CERN and Fermilab have measured the production of transverse energy, E_T, in 200-800 GeV proton-^{208}Pb collisions.[1] This data was analyzed[2] by assuming that the proton has independent interactions with each nucleon that it encounters. The high-energy nucleon-nucleon inelastic cross section is $\simeq$ 32 mb which corresponds to a disk of radius 1 fm. Therefore it is reasonable to assume that the proton will have an interaction with every nucleon within a cylinder of radius 1 fm, and will ignore the remaining nucleons. In this picture the transverse energy distribution resulting from a proton with impact parameter b is

$$\frac{1}{N}\frac{dN}{dE_T} = \sum_{n=1}^{A} P_n(b)\ F_n(E_T) \qquad (1)$$

*Work done with G. Baym, V. R. Pandharipande, and G. Friedman, University of Illinois, Urbana, IL where A is the number of nucleons in the nucleus, $P_n(b)$ is the probability of finding n nucleons in a cylinder of radius 1 fm and impact parameter b, and $F_n(E_T)$ is the transverse energy distribution arising from n independent nucleon-nucleon collisions. In Ref. 2, Poisson statistics were assumed for P_n; the present research is aimed at improving this assumption. The form of F_n is discussed later.

Consider the distribution of nucleons $P_n(V)$ in a subvolume, V, of a nucleus. The mean number, $\bar{n}$, is just the integral of the nucleon density over the volume V:

$$\bar{n} = \sum_n n\, P_n(V) = \int_V d^3r\, \rho(r)$$

$$= \int_V d^3r\, \langle\psi^+(r)\psi(r)\rangle\ . \tag{2}$$

In the last line we have written the number operator in terms of nucleon creation and destruction operators. The mean-square number fluctuation is

$$\sigma^2 = \langle(n-\bar{n})^2\rangle_V = \langle n^2\rangle_V - (\langle\bar{n}\rangle_V)^2 = \langle n^2\rangle_V - \bar{n}^2\ . \tag{3}$$

We can compute $\langle n^2\rangle_V$ by using the anticommutation rules for $\psi^+(r)$ and $\psi(r)$:

$$\langle n^2\rangle = \int_V d^3r_1 \int_V d^3r_2 \langle\psi^+(r_1)\psi(r_1)\psi^+(r_2)\psi(r_2)\rangle\ , \tag{4}$$

$$= \int d^3r_1 \langle\psi^+(r_1)\psi(r_1)\rangle$$

$$+ \int_V d^3r_1 \int_V d^3r_2 \langle\psi^+(r_1)\psi^+(r_2)\psi(r_2)\psi(r_1)\rangle\ , \tag{5}$$

$$= \bar{n} + \int_V d^3r_1 \int_V d^3r_2\, \rho(r_1)\rho(r_2)\, g_2(\vec{r}_1,\vec{r}_2)\ . \tag{6}$$

In the last line we have expressed the two-particle density in terms of the pair correlation function g_2. We can write Eq. (6) as

$$\langle n^2\rangle_V = \bar{n} + \bar{n}^2 + \int_V d^3r_1 \int_V d^3r_2 \; \rho(r_1)\rho(r_2) \left[g_2(\vec{r}_1,\vec{r}_2)-1\right] . \tag{7}$$

and thus

$$\sigma^2 = \bar{n} - \int_V d^3r_1 \int_V d^3r_2 \; \rho(r_1)\rho(r_2) \left[1-g_2(\vec{r}_1,\vec{r}_2)\right] . \tag{8}$$

Eq. (8) also holds for Bose statistics.

If Poisson statistics holds for P_n:

$$P_n(\text{Poisson}) = e^{-\bar{n}} \, \bar{n}^n/n! \; , \tag{9}$$

then we would have $\sigma^2=\bar{n}$ and the double integral of g_2-1 would have to be zero. One way of achieving this would be to have $g_2=1$, but this would imply no nucleon-nucleon correlations, not even correlations required by Fermi statistics. In fact from the known $g_2(r)$ for nuclear matter, it is clear that the integral in Eq. (8) must be significantly different from 0.

Before describing our results for P_n, I will give a brief overview of planned calculations of the ground states of finite nuclei. With V. R. Pandharipande and R. B. Wiringa, I am involved in the development of a program to compute the ground states of closed-shell nuclei. We intend to make variational calculations for a realistic Hamiltonian consisting of two-nucleon potentials, such as Argonne v_{14}, fitted to N-N phase shifts and three nucleon potentials, such as Urbana model VII, that give good ground states for both nuclear matter and A=3 and 4 nuclei. We are using a variational wave function of the form

$$\Psi = S \prod_{i<j} F_{ij} \, \Phi \,, \tag{10}$$

where S indicates the product over pairs is to be symmetrized and

$$\Phi = A \, |p\uparrow| \, |p\downarrow| \, |n\uparrow| \, |n\downarrow| \tag{11}$$

is a product of four determinants for the spin-up protons, spin-down protons, spin-up neutrons and spin-down neutrons. The antisymmetry operator, A, acts on the particles in different determinants. Each determinant contains the lowest A/4 eigenstates of a Woods-Saxon well.

The nucleon-nucleon correlations are

$$F_{ij} = f_c(r_{ij}) \, [1 + \sum_{p=2}^{8} u_p(r_{ij}) \, O^p_{ij}] \,, \tag{12}$$

where f_c and u_p are radial functions and the operator terms are

$$O^p_{ij} = \tau_i \bullet \tau_j, \; \sigma_i \bullet \sigma_j, \; \tau_i \bullet \tau_j \; \sigma_i \bullet \sigma_j \,,$$

$$S_{ij}, \; \tau_i \bullet \tau_j \; S_{ij}, \; (L \bullet S)_{ij}, \; \tau_i \bullet \tau_j \; (L \bullet S)_{ij} \,. \tag{13}$$

We will use

$$U_{ij} = \sum_p u_p(r_{ij}) \, O^p_{ij} \tag{14}$$

to designate the non-central parts of the nucleon-nucleon correlations.

Given the wave function of Eq. (10), we must evaluate

$$E = \frac{\langle \Psi | H | \Psi \rangle}{\langle \Psi | \Psi \rangle} \,. \tag{15}$$

We are using a cluster expansion in U_{ij} to do this. As an example, the expectation of the N-N potential is

$$\langle v_2 \rangle = \sum_{i<j} v_{ij} + \sum_{i<j<k} v_{ijk} + \dots \quad , \tag{16}$$

where v_{ij} consists of the contributions of just the two-body clusters:

$$v_{ij} = n_{ij}/(1 + d_{ij}) \quad , \tag{17}$$

$$n_{ij} = \frac{\langle \Psi_J | (1+U_{ij}) \; v(i,j) (1+U_{ij}) | \Psi_J \rangle}{\langle \Psi_J | \Psi_J \rangle} \quad , \tag{18}$$

$$= \int d^3r_1 \dots d^3r_A |\Psi_J|^2$$

$$\times \langle \Psi_J(r_1 \dots r_A) | (1+U_{ij}) v(i,j) (1+U_{ij}) | \Psi_J(r_1 \dots r_A) \rangle / |\Psi_J|^2$$

$$/ \int d^3r_1 \dots d^3r_A |\Psi_J|^2 \quad , \tag{19}$$

and

$$\Psi_J = \prod_{i<j} f_c(r_{ij}) \; \Phi \; . \tag{20}$$

The expression for d_{ij} is the same as that for v_{ij} with $v(i,j)$ replaced by unity.

The integral in Eq. (19) is evaluated by a Metropolis random walk controlled by $|\Psi_J|^2$. In this manner, the central part of the nucleon-nucleon correlation is treated exactly (up to Monte-Carlo statistical errors) and only the operator parts of the correlations are expanded.

A few comments on the computational aspects of this program are relevant. The determinants in Φ are small (4×4 for ^{16}O, 10×10 for ^{40}Ca) and hence their evaluation is not very efficient on vector processors. However, there are A(A-1)/2 (120 for ^{16}O or 780 for ^{40}Ca) pairs in the pair-correlation functions and much of the evaluation of these terms can be done with vectors whose length is the full A(A-

1)/2; this results in quite good speeds. The calculation of pair exchanges in Φ requires the evaluation of ratios of determinants such as (for ^{16}O)

$$|\vec{r}_5,\vec{r}_2,\vec{r}_3,\vec{r}_4|/|\vec{r}_1,\vec{r}_2,\vec{r}_3,\vec{r}_4| \ , \qquad (21)$$

in which the position of the fifth particle replaces that of the first particle. This is a simple update of one row of a determinant and, once the inverse of the original matrix is available, can be computed as one dot product. Furthermore, all of the updates can be evaluated in a single $[A \times \frac{A}{4}] \times [\frac{A}{4} \times A]$ matrix multiply which can be done very efficiently.

Our initial plan for the variational wave function is to use the pair correlation from nuclear matter, treating the density as a parameter. The parameters of the Woods-Saxon well for the single-particle wave functions will also be varied.

We used the random walks generated by this program to compute P_n in the following manner. Periodically during the walk, we counted the number of nucleons in cylinders of various radii and impact parameters. These counts were binned and then converted to the probabilities P_n. To get improved statistics, we used cylinders oriented along all three axes and for impact parameters $b>0$, we used several non-overlapping cylinders. We made enough samples to measure probabilities down to 10^{-6} and in some cases 10^{-8}.

The present program is limited to equal numbers of neutrons and protons. We made calculations for ^{40}Ca, but because much of the data are for ^{208}Pb we also considered the fictitious nucleus $^{224}112$. The parameters of the wave function were adjusted to give a reasonable density profile as shown in Fig. 1. To study the effects of the N-N correlations we made calculations with the correlated wave function of Eq. (20) (CWS), with a wave function containing

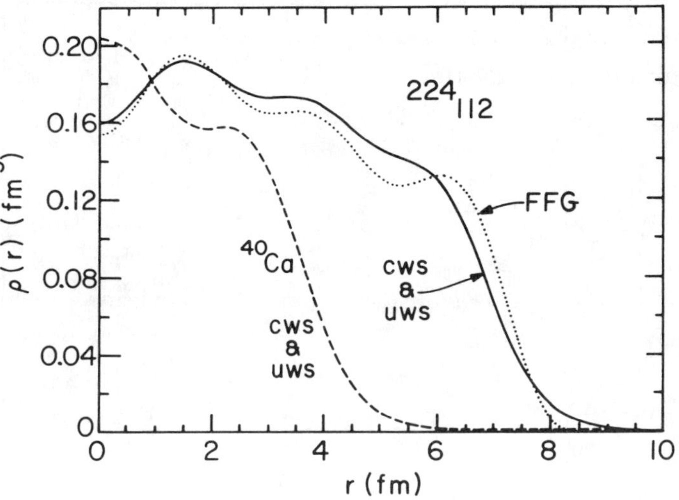

FIGURE 1. Density profiles for the nuclear wave functions used.

only the Φ (UCWS), and finally a wave function containing Slater determinants of spherical Bessel functions, $j_\ell(k_n r)$. The k_n were chosen such that

$$j_\ell(k_n R_0) = 0 \ , \tag{22}$$

with R_0 adjusted to give the desired rms radius for the nucleus. This last wave function is that of a finite Fermi gas (FFG).

In addition to the nuclear calculations, we made some calculations using ground-state wave functions for liquid drops of ^{3}He and ^{4}He atoms.[3] We considered systems of 40 and 240 ^{3}He atoms and 40, 240, and 728 ^{4}He atoms.

Figure 2 shows $\sigma^2/\bar{n}$ for cylinders of radius 1 fm passing through the nuclei at different impact parameters. The values are plotted against the $\bar{n}$ for each cylinder. For Poisson statistics, the ratio $\sigma^2/n=1$. As can be seen, the values fall into a band dropping to about 0.7 for the largest

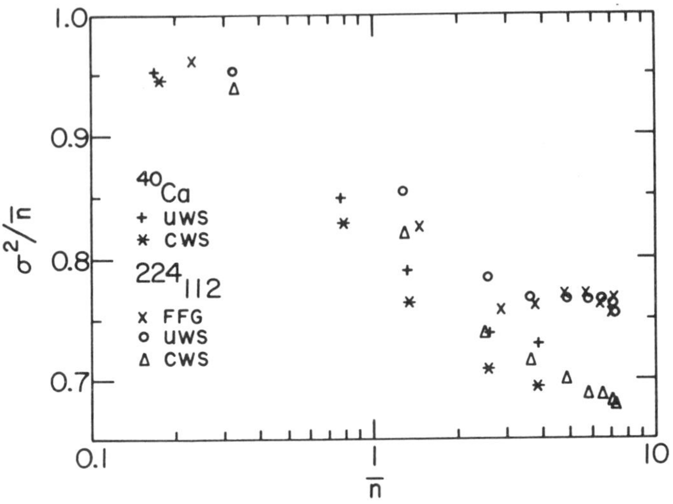

FIGURE 2. Values of $\sigma^2/\bar{n}$ versus $\bar{n}$ for cylinders of radius 1 fm through nuclei.

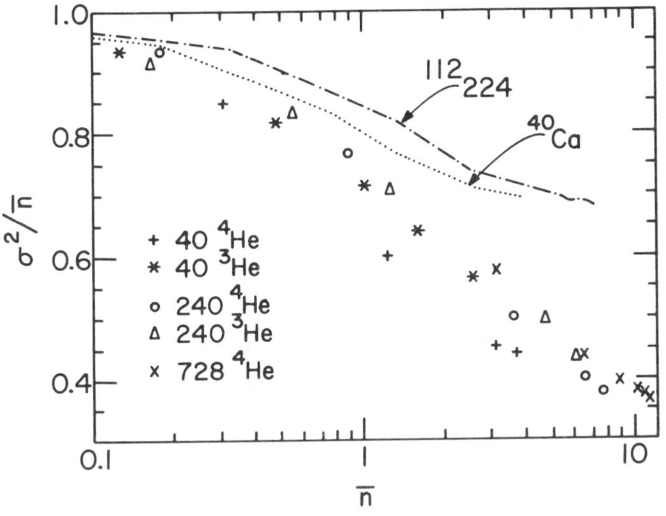

FIGURE 3. Values of $\sigma^2/\bar{n}$ versus $\bar{n}$ for cylinders of radius 2 Å through liquid drops of ^{3}He and ^{4}He. The curves show the CWS nuclear results for cylinders of radius 1 fm.

$\bar{n}$ possible in these cases. The values for the uncorrelated wave functions (FFG and UWS) are somewhat above these for the CWS cases. There is only a small nuclear size effect. Figure 3 shows similar results for cylinders of radius 0.8 σ (2.0 Å) passing through the helium drops. Again there is a narrow band containing both the ^{3}He and ^{4}He results; in this case most of the correlations come from the very strong repulsive potential in the He-He interaction. The He-He correlations are stronger than those for nucleons, and this $\sigma^2/\bar{n}$ for a given $\bar{n}$ is lower than in the nuclei.

Figures 4 and 5 show P_n for 1 fm radius cylinders with several impact parameters. The curves show the following probability distributions:

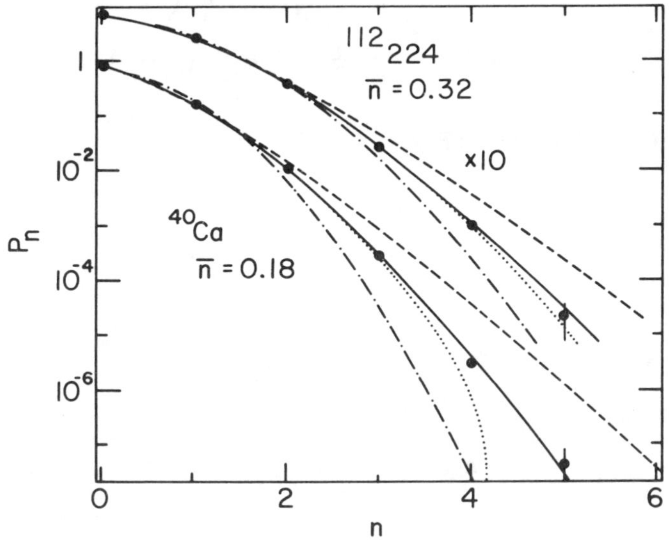

FIGURE 4. P_n for cylinders of radius 1 fm through nuclei with CWS wave functions. See the text for a description of the curves.

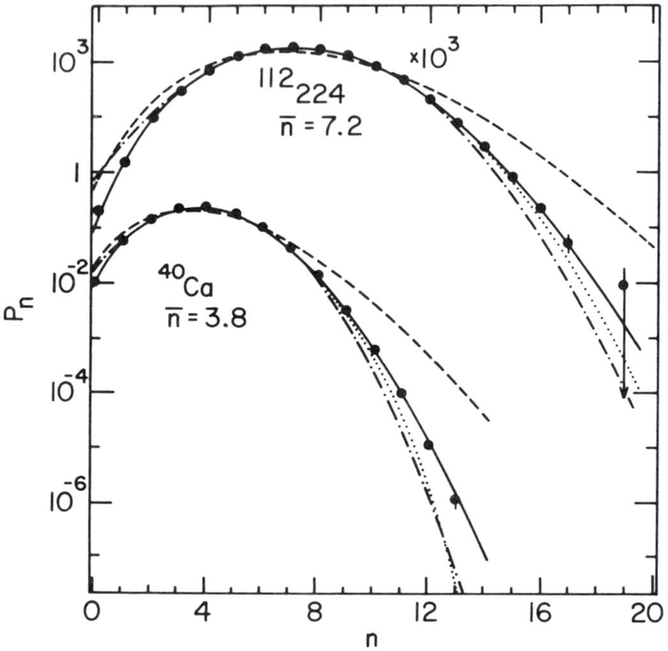

FIGURE 5. See Fig. 4.

Poisson (dashed)

$$P_n = e^{-\bar{n}} \frac{\bar{n}^n}{n!} \quad ; \tag{23}$$

Binomial (dotted)

$$\begin{aligned} P_n &= B\ (1-a)^{N-n}\ a^n\ \frac{\Gamma(N+1)}{n!\Gamma(N-n+1)}\ , \quad n \le N\ , \\ &= 0\ , \qquad\qquad n > N\ ; \end{aligned} \tag{24}$$

Normal (dash-dot)

$$P_n = C\ e^{-\frac{1}{2}\left(\frac{c-n}{d}\right)^2} \quad ; \tag{25}$$

Reducing Poisson (solid)

$$P_n = D \frac{\alpha^{n(n-1)/2} \, p^n}{n!} \quad . \tag{26}$$

In all cases the parameters were adjusted to reproduce the $\bar{n}$ and σ^2 of the Monte Carlo values (for Poisson only $\bar{n}$ is fit).

As can be seen, the Poisson distributions are always too broad. Because the values of N in the binomial distribution fits usually turned out to be rather small, these distributions often fall to zero at values of n for which the Monte Carlo $P_n > 0$. The normal distribution works well only for cases with large $\bar{n}$.

As can be seen from the solid curves, the reducing Poisson distributions are always a good representation of the Monte Carlo P_n. The standard Poisson distribution is obtained by assuming that there is a large number of nucleons available to put in the cylinder, and that the probability of any one being in the cylinder is independent of how many others are there. The reducing Poisson distribution is obtained by assuming that this probability is proportional to α^i where i is the number of nucleons already in the cylinder. Because $\alpha<1$, this quite naturally represents the fact that the presence of some nucleons tends to exclude other nucleons.

The P_n for the uncorrelated nucleons (UWS and FFG) and the helium drops have the same features. The reducing Poisson is always a good distribution and the other distributions fail in the same ways.

I now return to the calculation of the E_T distribution. To compute $F_n(E_T)$, we assume that each N-N collision is independent and thus F_n is a folding of F_1:

$$F_n(E_T) = \int_0^\infty dE_1 \int_0^\infty dE_2 \cdots \int_0^\infty dE_n \; \delta(E_T - E_1 - E_2 - \cdots - E_n) \times$$

$$F_1(E_1)F_1(E_2)\dots F_1(E_n) \quad . \tag{27}$$

$F_1(E_T)$ is measured[4] in pp collisions and is well represented by an exponential distribution. To allow us to study the effects of different F_1, we consider a more general form:

$$F_1(E_T) = \frac{1}{\Gamma(\gamma)} \left(\frac{\gamma\, E_T}{\epsilon}\right)^{\gamma-1} e^{-\gamma\, E_T/\epsilon} \quad . \tag{28}$$

Here ϵ is mean value of E_T and

$$\gamma = \epsilon^2/\sigma^2(E_T) \tag{29}$$

is a parameter that allows us to adjust the width of the $F_1(E_T)$ distribution. The experimentally observed $F_1(E_T)$ correspond to γ=1. Figure 6 shows $F_1(E_T)$ for ϵ=1.8 GeV and γ=1, 0.75, 0.5. As is explained in Ref. 2, ϵ=1.8 GeV is a reasonable value that takes into effect rescattering in a

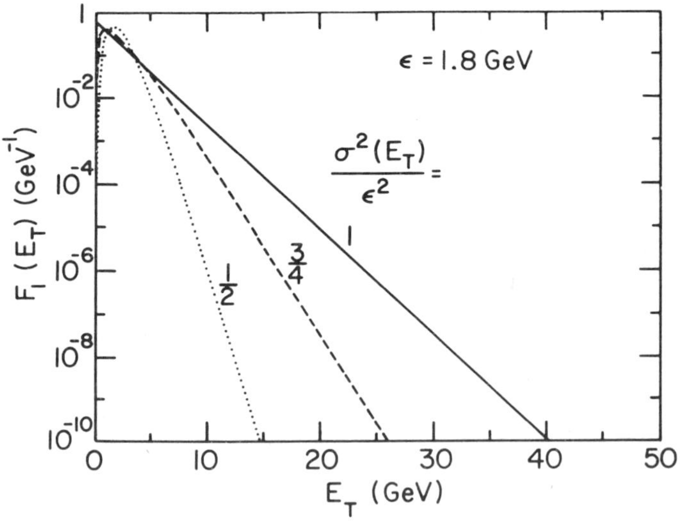

FIGURE 6. Distributions of E_T produced in N-N collisions from Eq. (28) with ϵ=1.8 GeV.

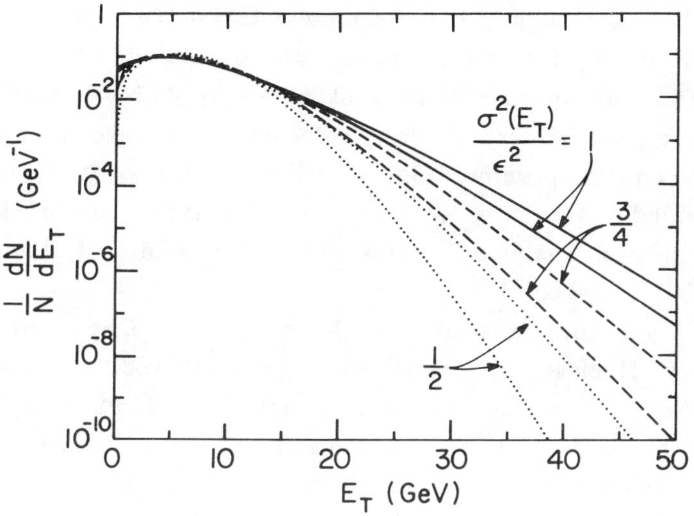

FIGURE 7. Distributions of E_T for a proton passing through a cylinder with $\bar{n}=3.6$. The curves are explained in the text.

^{208}Pb nucleus. Figure 7 shows the computed E_T distributions using the $F_1(E_T)$ of Fig. 6 and P_n for $\bar{n}=3.6$ (this corresponds to the average over impact parameters for p-^{208}Pb collisions). For each pair of curves, the upper curve is obtained using the Poisson distribution for P_n while the lower is based on the reducing Poisson distribution. The solid curves correspond to $\gamma=1$. As can be seen, the strong difference between the Poisson and reducing Poisson distribution for P_n (Fig. 5) is largely eliminated by the broad $F_1(E_T)$ distribution determined by the pp data; the difference between the two solid curves is probably not experimentally observable. The sharper distributions for $F_1(E_T)$ given by the dashed and dotted curves in Fig. 6 do, of course, result in correspondingly larger differences in the E_T distributions shown in Fig. 7.

In conclusion, we have demonstrated that the number distribution of nucleons in a sub-volume of a nucleus is substantially sharper than that given by Poisson statistics. However, this difference does not seemed to have a practical effect on the predicted E_T distribution for proton-nucleus collisions. We have also studied the expansion of P_n in powers of n analytically; this and other results of this work will be published elsewhere.

This work is supported by the U. S. Department of Energy, Nuclear Physics Division, contract W-31-109-ENG-38 and Materials Science Division, contract DE-AC02-76ER01198, and by the National Science Foundation, Grant PHY84-15064. The calculations were made possible by a grant of computer time on the Energy Research Cray of the U. S. Department of Energy.

REFERENCES

1. Åkesson et al., Nucl. Phys. A **447**, 475c (1985); contribution to 23rd Intern. Conf. on High Energy Physics (Berkeley, July 1986); Fermilab E-609 Collab., J. A. Rice et al., Nucl. Phys. A **418**, 315c (1984); E-557 Collab., R. Gomez et al, preprint FERMILAB-Conf. 86/85-E.
2. G. Baym, P. Braun-Munzinger and V. Ruuskanen, Phys. Lett. B **190**, 29 (1987).
3. V. R. Pandharipande, S. C. Pieper and R. B. Wiringa, Phys. Rev. B **34**, 4571 (1986).
4. C. Bromberg et al., Nucl. Phys. B **107**, 82 (1976).

NUMERICAL SOLUTION OF HIGH TEMPERATURE SUPERCONDUCTOR SPIN SYSTEMS

T. Barnes

Department of Physics
University of Toronto
60 St. George St.
Toronto, Ontario M5S 1A7
Canada

Invited contribution to

"Computational Atomic and Nuclear Physics at One Gigaflop"

Oak Ridge, Tennessee, 14-16 April 1988.

Abstract

Recent experimental work on high temperature superconductors suggests that these materials may be regarded as two dimensional antiferromagnetic quantum spin systems. This has stimulated intense theoretical interest in the two dimensional Hubbard and Heisenberg models, which are thought to describe these systems in certain limits. In this paper we describe a numerical study of the two dimensional Heisenberg model using a Hamiltonian Monte Carlo algorithm which simulates the Euclidean time Schrödinger equation using random walks. This algorithm allows the numerical measurement of energies and expectation values of the ground state of a specified Hamiltonian. We also discuss the optimization of the algorithm and show that energy measurement errors can be made arbitrarily small with sufficiently well chosen importance sampling.

HIGH-T_c SUPERCONDUCTIVITY AND SPIN SYSTEMS

An unconventional mechanism underlies high-T_c superconductivity

In 1986 when Bednorz and Müller discovered high temperature superconductivity in barium-doped La_2CuO_4 [1], they suggested that this material was a conventional BCS superconductor, and that the high critical temperature was due to a large electron-phonon coupling. This is no longer believed to be the case, and the search for the mechanism which produces superconductivity in these new materials is the central theoretical problem in this field. For purposes of orientation we shall first recall some results from the theory of conventional superconductivity and note why this theory is inappropriate for the new materials. Following this we will briefly discuss experimental results which have suggested two dimensional spin systems as appropriate models for high temperature superconductivity.

The mechanism of conventional superconductivity was first studied in detail by Bardeen, Cooper and Schrieffer in 1957 [2], who noted that the many-fermion ground state of an ideal metal was energetically unstable with respect to e^-e^- pair formation, given an attractive interaction. This "Cooper pair" description of the superconducting state successfully explained a number of well established features of superconductivity, such as the gap in the energy spectrum (seen in the specific heat and in superconductor-normal metal tunnelling), the coherence length ξ_0 (the approximate size of a Cooper pair, which follows from the uncertainty principle), the critical temperature T_c, the electromagnetic penetration depth $\lambda(T)$, and many related quantities. The critical temperature was determined in a variational calculation using a simple trial ground state having a spin-singlet pairing amplitude and a model phonon-exchange Hamiltonian, with a potential depth of V_0 for electrons having energies within $\hbar\omega_D$ of the Fermi energy E_F. Here ω_D is the Debye frequency, which is an approximate ultraviolet cutoff frequency for available phonon modes. The BCS result relating T_c, ω_D and the energy gap $2\Delta(0)$ is

$$3.53\, k_B T_c = 2\Delta(0) = 4\hbar\omega_D \exp(-1/N(0)V_0) \tag{1.1}$$

where k_B is Boltzmann's constant and N(0) is the density of states at the Fermi surface.

In principle the BCS formalism applies to any attractive electron-electron interaction; convincing evidence in favor of phonon exchange as the interaction underlying normal superconductivity is the "isotope effect". As phonons represent states of vibration of the electron-ion lattice, the maximum possible phonon frequency (which is close to ω_D) is $\sqrt{k/M_{ion}}$, assuming SHO restoring forces. This implies that the critical temperature of a single-element superconductor depends on the isotopic mass through (1.1) as

$$T_c = k\, M^{-1/2} \ . \tag{1.2}$$

In many ways the high temperature superconductors appear to be BCS-type systems. They have an energy gap which is perhaps a factor of two larger than 3.53 $k_B T_c$ (1.1), charge carriers with $|Q| \approx |2e|$, a coherence length ξ_0 of about 10-20 Å, $\sim 10 - 10^2$ times smaller than for conventional superconductors, as expected if $\xi_0 \sim \hbar/\delta p \sim \hbar v_F/2\Delta(0) \propto T_c^{-1}$, and a penetration depth $\lambda(T)$ with a temperature dependence very similar to the BCS prediction. (In view of the vast number of experimental references, in this introduction I simply refer to a conference report [3] which gives details of many experimental results.)

Problems with the conventional BCS picture include Hall effect measurements which suggest that the dominant charge carriers are holes rather than electrons, specific heat measurements which see an approximately linear T dependence in $c_v(T)$ at intermediate temperatures rather than the expected $\exp(-E_{gap}/k_B T)$, and the lack of an isotope effect (1.2) in the critical temperature. The absence of a significant isotope effect in the new materials is the strongest evidence against phonon exchange as the pairing mechanism, and has led to many other proposals for interactions which might cause electron or hole pairing.

High-T_c superconductors as two dimensional spin systems

A comparison of the unit cells of the known high temperature superconductors shows that their common feature is CuO_2 planes. (Note added in proof: a material having the formula $Ba_{0.6}K_{0.4}BiO_3$ with a T_c of about $30^o K$ which has a related perovskite structure but is free of copper has been discovered very recently [4]. There are indications that the mechanism of superconductivity in this material is distinct from the spin system models of copper-oxygen materials discussed here [5].)

One of these CuO_2 planes is visible in the unit cell of La_2CuO_4 (the first high temperature superconductor) at the top of figure 1 [6]; the $\sim$ 100°K materials found more recently have multiple planes. The problem of the origin of high temperature superconductivity thus reduces to one of understanding the physics of electrons in these two dimensional CuO_2 systems. One possible resolution of the difficulties with the BCS picture noted above is to suggest that pairing of holes rather than electrons is taking place in the CuO_2 planes, and that the pairing mechanism is distinct from phonon exchange.

A Hamiltonian often proposed to describe the behavior of electrons and holes in the CuO_2 planes is the Hubbard Hamiltonian, which has a "hopping term" to allow electron motion and an on-site Coulomb repulsion U which discourages double occupancy;

$$H_{Hubbard} = -t \sum_{<ij>,\,\sigma} c^{\dagger}_{i\sigma} c_{j\sigma} + U \sum_{i} n_{i\uparrow} n_{i\downarrow} \ . \qquad (1.3)$$

This model of the undoped "precursor insulators" such as La_2CuO_4 anticipates antiferromagnetic alignment of the adjacent electrons in the CuO_2 planes, because singlet states such as $|\uparrow\downarrow\rangle$ repel doubly occupied states, for example $|\uparrow\downarrow \ \cdot\rangle$, and hence are driven lower in energy than the triplet $|\uparrow\uparrow\rangle$ and $|\downarrow\downarrow\rangle$ states. This antiferromagnetic alignment has been observed experimentally [7].

In the limit of large on-site repulsion ($U/t >> 1$, believed to be correct for high-T_c), assuming one electron per site (a "half-filled band"), the Hubbard model reduces to the more tractable Heisenberg antiferromagnet, with the Hamiltonian

$$H_{Heisenberg} = \frac{4t^2}{U} \sum_{<ij>} \vec{S}_i \cdot \vec{S}_j \ . \qquad (1.4)$$

Additional support for a mechanism relating spin systems and superconductivity comes from behavior of the superconductors as a function of the amount of hole doping, for example in $La_{2-x}Sr_xCuO_4$ [8]. At $x = 0$ this material is free of holes and is an insulator, with a single conduction electron on each of the copper sites in the planes. The conduction electrons are held on Cu sites by their mutual Coulomb repulsion, which makes the material a "Mott-Hubbard insulator" rather than a metal; metallic behavior would naively be expected given the presence of the unpaired electrons on the Cu^{++} sites. As the doping fraction in the insulator is increased the amount of long range electron-spin antialignment (the "staggered magnetization") decreases, and vanishes altogether at a critical doping fraction of $x_c \approx 0.05$. Immediately above x_c the material is a superconductor. This dramatic correlation between long range spin order, hole doping and superconductivity has given much encouragement to studies of the two dimensional Hubbard and Heisenberg models.

Since superconductivity is known to arise when the amount of hole doping in the otherwise half-filled band exceeds $x_c \approx 0.05$, numerical studies of the Hubbard and Heisenberg models near half filling may lead to an understanding of the precise mechanism underlying high temperature superconductivity. As these models are hard to solve numerically, many of the mechanisms proposed for high temperature superconductivity (see for example [9-17]) are simply speculations about the properties of the ground state of (1.3) or (1.4). The situation is not unlike that in QCD, which is studied using approximate models although the correct Hamiltonian is known. In both fields the extraction of predictions directly from the proposed theory is a difficult problem which is most suitably addressed using Monte Carlo algorithms.

THE ALGORITHM AND APPLICATION TO HIGH-T_c SPIN SYSTEMS

Quantum matrix elements from the Euclidean time Schrödinger equation

The general problem we wish to solve is to determine matrix elements $\langle\psi_0|Q|\psi_0\rangle$ for the ground state and low lying excited states of a system having a specified Hamiltonian H. We assume that the Hilbert space is so large that direct diagonalization of the Hamiltonian matrix is not practical, so that we must instead use a stochastic method. Our approach is to set up a Monte Carlo simulation of the Schrödinger equation in Euclidean time, which is

$$-\frac{\partial}{\partial\tau}\,|\psi(\tau)\rangle = H|\psi(\tau)\rangle\,. \tag{2.1}$$

This Euclidean time state vector is of interest because it converges to the ground state wavefunction as the Euclidean time (τ) goes to infinity, as is evident from an eigenmode expansion;

$$|\psi(\tau)\rangle = \sum_n c_n|\psi_n\rangle\, e^{-E_n\tau}\ ; \tag{2.2}$$

$$\lim_{\tau\to\infty}\ |\psi(\tau)\rangle = c_0\, e^{-E_0\tau}\Big(\,|\psi_0\rangle\ + O(e^{-E_{gap}\tau})\Big)\,. \tag{2.3}$$

Evidently one may recover the ground state $|\psi_0\rangle$ and the ground state energy E_0 from the large τ behavior of the Euclidean time wavefunction, provided that the initial distribution $|\psi(0)\rangle$ is not orthogonal to the ground state,

$$c_0 = \langle\psi_0|\psi(0)\rangle \neq 0\,. \tag{2.4}$$

Generating ground state properties by simulating the Euclidean time Schrödinger equation has been exploited in several algorithms, for example the Green's Function Monte Carlo method [18] and the Projector Monte Carlo technique of Hirsch *et al* [19]. Various related methods are discussed in the text of Kalos and Whitlock [20]. All these methods attempt to solve (2.1) by simulating an equivalent diffusion process in real time. The method we employ is very similar to the Projector Monte Carlo method, the principal difference being the use of a fixed basis, whereas PMC employs a rapidly alternating one. Our technique has been described in a number of previous references [21-25], although some of the results described here, in particular the optimization of the discrete algorithm, have not appeared elsewhere.

The random walk algorithm for continuous Hilbert spaces

We simulate the Euclidean time evolution of the Schrödinger equation by establishing a relation between it and a diffusion process in real time, which may be modelled on a computer. To illustrate this in a familiar context, compare the one dimensional Schrödinger equation in Euclidean time with the diffusion equation in real time;

$$-\frac{\partial}{\partial \tau}\, \psi(x,\tau) = -\frac{1}{2m}\, \nabla^2 \psi(x,\tau) \;+\; V(x)\, \psi(x,\tau)\,, \tag{2.5}$$

$$-\frac{\partial}{\partial t}\, \rho(x,t) = -\kappa\, \nabla^2 \rho(x,t)\,. \tag{2.6}$$

These are equivalent in the free case, $V(x) = 0$. As we are interested in problems with nontrivial potential terms, we generalize (2.6) to diffusion in the presence of a static distribution of absorbers, $a(x)$. This gives a diffusion equation with sources,

$$-\frac{\partial}{\partial t}\, \rho(x,t) = -\kappa\, \nabla^2 \rho(x,t) + a(x)\, \rho(x,t)\,, \tag{2.7}$$

which is equivalent to the Euclidean time Schrödinger equation provided that we identify t with τ, κ with $1/2m$, $a(x)$ with $V(x)$ and the density of particles $\rho(x,t)$ with the wavefunction $\psi(x,\tau)$. So, to obtain ground state properties of a quantum mechanical system we need only simulate the diffusion problem described by (2.7). (We will henceforth use τ to represent both Euclidean time in the Schrödinger equation and real time in the equivalent diffusion problem.)

We model this diffusion process by generating a sequence of random walks on a lattice in (x,τ) space. As an example of this procedure we first consider the free diffusion process (2.6). If we start a random walk at the origin ($x = 0$ at $\tau=0$) and require that it step from x to either of the two sites $x \pm h_x$ with equal probability between τ and $\tau + h_\tau$, we obtain a recursion relation for the probability of occupation $p(x,\tau)$, which is

$$p(x,\tau + h_\tau) = \frac{1}{2}\Big(p(x - h_x,\tau) + p(x + h_x,\tau)\Big)\,. \tag{2.8}$$

This probability of occupation is of course proportional to the density of particles $\rho(x,\tau)$ in the diffusion problem. If the lattice step sizes h_x and h_τ are small relative to the spatial and temporal variation of p, we recover the diffusion equation

$$-\frac{\partial}{\partial \tau}\, p(x,\tau) = -\frac{h_x^2}{2h_\tau}\, \nabla^2 p(x,\tau)\,. \tag{2.9}$$

This is both the diffusion equation (2.6) in real time and the free Schrödinger equation in Euclidean time if we relate the lattice step sizes, the diffusion coefficient and the mass by

$$\frac{h_x^2}{2h_\tau} = \kappa = \frac{1}{2m} \ . \tag{2.10}$$

Thus, a histogram of random walk arrival points $\{x\}$ at time τ gives an estimate of $p(x,\tau)$ and hence $\psi(x,\tau)$ which becomes exact as the lattice spacing h_x is taken to zero and the number N_{rw} of walks generated goes to infinity.

In the presence of a potential $V(x)$ we must simulate (2.7) rather than (2.6). We do this by constructing a histogram of weight factors $\{w^{(m)}\}$, where the weight $w^{(m)}$ is the probability that the mth walk would have escaped absorption on the path $x^{(m)}(\tau)$ it followed, given the presence of the distribution of absorbers $a(x) = V(x)$. This survival probability is

$$P^{(m)}_{survive}\Big((0,0) \to (x,\tau)\Big) = \exp\Big(-\int_0^\tau V(x^{(m)}(\tau'))\,d\tau'\Big) \equiv w^{(m)} \ , \tag{2.11}$$

so a histogram of the weights $\{w^{(m)}\}$ generated by the N_{rw} walks ($m = 1,...,N_{rw}$) satisfies the diffusion equation with sources (2.7) and hence the Euclidean Schrödinger equation (2.5). This result, which gives the quantum amplitude to go from $(0,0)$ to (x,τ) as an average of (2.11) over random walks, may be concisely written as a generalization of the usual path integral;

$$A(i \to f) = \sum_{\substack{paths \\ i\to f}} \exp(-\mathbf{S}) = \Big\langle \exp(-\mathbf{V})\Big\rangle_{\substack{random\ walks \\ i\to f}} \tag{2.12}$$

where

$$\mathbf{V} \equiv \int_{\tau_i}^{\tau_f} V(x(\tau))\,d\tau \ . \tag{2.13}$$

Energies may be obtained from the time evolution of the mean (random walk averaged) weight $<\omega(\tau)>$ (2.11), since the identification of the weight factor histogram with the wavefunction $\psi(x,\tau)$ implies that the mean weight asymptotically decays as $\exp(-E_0\tau)$ at large τ. One may also extract matrix elements from a two stage walk [21], or from the Feynman-Hellman theorem, as will be described at the end of the section on applications.

In practice the simple result represented by (2.11-13) is seldom useful, because problems of interest typically have potentials that take on a wide range of values. Given such a potential, unguided random walks generally wander into classically forbidden regions where the potential $V(x)$ is large, which leads to a large scatter of the weight factors $\{w^{(m)}\}$ and hence to large statistical errors. For this reason it is useful to introduce importance sampling in the algorithm, through a bias in the choice of path followed by a walk. In the one dimensional example, instead of giving the steps $(x,\tau) \to (x \pm h_x, \tau + h_\tau)$ equal probability, we instead bias the stepping probability according to

$$p(x,\tau \to x \pm h_x, \tau + h_\tau) = \frac{1}{2} \mp h_x \, \frac{\partial g(x)}{\partial x} \, . \tag{2.14}$$

We call $g(x)$ the "guiding potential". This generalization preferentially guides walks towards minima of $g(x)$, and at large τ the walks approach a probability distribution

$$\lim_{\tau \to \infty} p(x,\tau) = k \; e^{-4g(x)} \tag{2.15}$$

rather than the flat distribution which results from uncontrolled diffusion. To recover the Schrödinger equation from a weight factor histogram with the path importance sampling (2.14) we must generalize the simple weight factor $\exp(-\int V(x)d\tau)$ in (2.11) to the more complicated form

$$w^{(m)} = w^{(m)}_{endpoint} \cdot w^{(m)}_{path} \; ; \tag{2.16}$$

$$w_{endpoint} = e^{2g(x(\tau))} \; , \tag{2.17}$$

$$w_{path} = \exp\left(- \int_0^T \left(V(x(\tau')) + \nabla^2 g(x(\tau')) - 2(\vec{\nabla} g(x(\tau')))^2 \right) d\tau' \right) . \tag{2.18}$$

As a histogram of these weights in x at time τ satisfies the Euclidean time Schrödinger equation, it may be identified with the wavefunction $\psi(x,\tau)$. (We have specialized to unit mass in (2.18).)

A crucial simplification of the algorithm is possible if we wish to measure energies alone and not matrix elements. We may set the "endpoint weight" $\exp\{2g(x(\tau))\}$ in (2.16-17) above equal to unity, as a histogram of the "path weights" alone gives the function

$$e^{-2g(x)} \; \psi(x,\tau) \; , \tag{2.19}$$

which has the same $e^{-E_0\tau}$ time dependence at large Euclidean time. This procedure usually gives a smaller scatter of weight factors and hence a more accurate estimate of the energy.

To optimize the algorithm we wish to make the variance of the weight factors $\{w^{(m)}\}$ in (2.16) a minimum with respect to $g(x)$. If we measure energies only we discard the endpoint weights as noted above and try to minimize the variance of the path weights (2.18). By inspection a perfect (nonsingular) solution for $g(x)$ would make all the path weights equal to the same path-independent constant, and hence would satisfy

$$V(x) + \nabla^2 g(x) - 2\big(\vec{\nabla} g(x)\big)^2 = constant \equiv E_0 \ . \tag{2.20}$$

For such a guiding potential the statistical error in E_0 measurement using $< w(\tau) >$ would be zero; a single walk would suffice to measure E_0 exactly! This "perfect guidance" $g^p(x)$ is given by

$$g^p(x) = -\frac{1}{2} \ln \psi_0(x) \ . \tag{2.21}$$

Of course we generally do not know ψ_0, and so must use a parametrized Ansatz for $g(x)$, or equivalently for a trial ground state wavefunction, $\psi_0^g(x) \equiv \exp(-2g(x))$. The optimum parameters in the $g(x)$ Ansatz may then be determined by minimizing the statistical error of the quantity being measured. A convenient measure of the scatter of weight factors to be minimized is the "dispersion" of w, which we define by

$$\mathrm{disp}(w(\tau)) \equiv \frac{\sigma_w(\tau)}{< w(\tau) >} \ . \tag{2.22}$$

We emphasize that the algorithm converges to correct energies and matrix elements for *any* $g(x)$ which has correct nodes; a choice close to $g^p(x)$ merely serves to reduce the number of walks required to attain a given statistical accuracy.

The random walk algorithm for discrete Hilbert spaces

The algorithm we have devised for generating ground state properties of systems having a discrete set of basis states is closely analogous to the continuum algorithm reviewed above. We assume that a Hamiltonian H and a complete set of basis states $\{|s\rangle\}$ have been specified, and we again study the large τ limit of the Euclidean time Schrödinger equation

$$-\frac{\partial}{\partial \tau} |\psi(\tau)\rangle = H|\psi(\tau)\rangle \tag{2.23}$$

by simulating an equivalent diffusion process in the configuration space $\{s\}$. As in the quantum mechanical problem the one-to-one correspondence between classical configurations x and basis states $|x\rangle$ (here s and $|s\rangle$) makes possible an isomorphism between classical diffusion in real time and quantum dynamics in Euclidean time. For the spin systems considered here the labels $\{s\}$ specify S_z-diagonal spin configurations, for example in the $N = 2$ one dimensional spin system s can assume the values $\uparrow\uparrow$, $\uparrow\downarrow$, $\downarrow\uparrow$ and $\downarrow\downarrow$.

The classical diffusion problem we model is the following: assume that the diffusion is described by a random walk through configuration space, and that if the walk is at a configuration s at time τ it may step to one of the "nearest neighbor" configurations $s' \neq s$ at time $\tau + h_\tau$ with a probability of

$$p(s,\tau \rightarrow s',\tau + h_\tau) = r_{ss'} h_\tau \ . \tag{2.24}$$

The walk may also remain at the configuration s in the interval $\tau \rightarrow \tau + h_\tau$ with a probability of $1 - \sum_{s'} r_{ss'} h_\tau$. The transition rates $\{r_{ss'}\}$ are a set of arbitrary positive numbers, provided that the $\{p(s \rightarrow s')\}$ in (2.24) satisfy the obvious constraints on probabilities. "Nearest neighbor" configurations are defined to be those pairs (s, s') for which the matrix element of H_I (the off-diagonal part of H) is nonzero,

$$\langle s'|H_I|s\rangle \neq 0 \ \text{ for } \ (s,s') = \ \text{nearest neighbors} \ . \tag{2.25}$$

The $\{r_{ss'}\}$ form a "guiding matrix" which biases the path followed through configuration space in a manner analogous to $g'(x)$ in the quantum mechanical algorithm.

The remaining, diagonal part of H is H_0, whose matrix elements define a potential $V_0(s)$,

$$H_0 = \sum_s V_0(s)\ |s\rangle\langle s| \ . \tag{2.26}$$

To generate a solution of the Euclidean Schrödinger equation we again construct a histogram of weight factors in the configuration space of the classical problem. In this case the weight factor associated with the mth random walk is the product of two factors, an exponential "diagonal weight" which depends on the path followed in configuration space and a "transition weight", which is the product of N_T factors, where N_T is the number of spin-state transitions which have taken place during the walk;

$$w_{total}^{(m\text{th walk})}(0 \rightarrow \tau) = w_{transition}^{(m)} \cdot w_{diagonal}^{(m)} \ ; \tag{2.27}$$

$$w_{transition} = \prod_{\substack{s \rightarrow s' \\ transitions}} \left(- \frac{\langle s'|H_I|s\rangle}{r_{ss'}} \right) , \tag{2.28}$$

$$w_{diagonal} = \exp\left(- \int_0^\tau \left(V_0(s(\tau')) - \sum_{s'} r_{ss'} \right) d\tau' \right) . \tag{2.29}$$

This discrete algorithm actually contains the continuum algorithm as a special case. For example, in the unguided one dimensional quantum mechanical problem of (2.8-11) we have

$$p\Big(x,\tau \to x \pm h_x, \tau + h_\tau\Big) \equiv r_{x,x\pm h_x} h_\tau = 1/2 \tag{2.30}$$

and

$$\langle s'|H_I|s\rangle \equiv \langle x'| \Big(-\frac{1}{2m}\nabla^2\Big) |x\rangle \Big|_{x'\neq x} = -\frac{1}{2mh_x^2}\Big(\delta_{x',x-h_x} + \delta_{x',x+h_x}\Big) \,. \tag{2.31}$$

Comparison with (2.25) shows that x and $x \pm h_x$ are nearest neighbor configurations in the sense defined by the discrete algorithm, and the transition weight (2.28) due to a single transition $x, \tau \to x + h_x, \tau + h_\tau$ is

$$w_{transition} = -\frac{\langle x + h_x|H_I|x\rangle}{r_{x,x+h_x}} = \frac{\frac{1}{2mh_x^2}}{\frac{1}{2h_\tau}} = \frac{h_\tau}{mh_x^2} \,. \tag{2.32}$$

Specializing to $h_\tau = mh_x^2$ is nonessential but simplifies the algorithm by making the transition weight (2.32) in (2.27) unity. This was *required* by the continuum algorithm in (2.10). Given this constraint, the integrand in the diagonal weight (2.29) becomes

$$V_0(x) - \sum_{x'} r_{xx'} = \langle x| \Big(-\frac{1}{2m}\nabla^2 + V(x)\Big) |x\rangle - r_{x,x-h_x} - r_{x,x+h_x}$$

$$= \frac{1}{mh_x^2} + V(x) - \frac{1}{h_\tau} = V(x) \,, \tag{2.33}$$

so we recover the unguided continuum weight factor (2.11),

$$w^{(m)} = \exp\Big(- \int_0^T V(x^{(m)}(\tau'))\, d\tau'\Big) \tag{2.34}$$

as a special case of the discrete algorithm. We may similarly recover the guided continuum weight factor (2.16) from (2.27) as a special case with $h_\tau = mh_x^2$.

Recall that one is able to optimize the continuum algorithm provided that one can find the "perfect" guiding potential $g^p(x) = -\ln(\psi_0)/2$, and that this choice allows exact energy measurements from a single walk. We naturally enquire whether there exists an analogous "perfect" choice for the guiding matrix $r_{ss'}$. The answer is in the affirmative, and the perfect $r_{ss'}$ is

$$r^p_{ss'} = -\langle s'|H_I|s\rangle \frac{\psi_0(s')}{\psi_0(s)} \,. \tag{2.35}$$

(For convenience we assume all the nonzero matrix elements of H_I are negative, which follows with a suitable choice of phases for the problems considered here.) To obtain energy measurements with zero statistical error given (2.35) one must discard the transition weight factor in (2.27) and calculate energies using only the diagonal weight (2.29). This simplification is justified for any $r_{ss'}$ of the general form

$$r_{ss'} = - \langle s'|H_I|s\rangle \, \frac{\psi_0^g(s')}{\psi_0^g(s)} \tag{2.36}$$

because the transition weight is a function of the arrival configuration $s(\tau)$ only and has no explicit time dependence. This ψ_0^g may be regarded as a parametrized trial Ansatz for ψ_0, analogous to the $\psi_0^g \equiv \exp(-2g)$ used for path importance sampling in the continuum algorithm. Substitution into (2.28) shows that with the form (2.36) the transition weight is a function of $s(\tau)$ only; since $p(s,\tau)$ has a limiting distribution $p(s,\infty)$ at large τ, the asymptotic time dependence $e^{-E_0\tau}$ evidently arises exclusively from the diagonal weight. This argument is analogous to the one which justified discarding the endpoint weight factor e^{2g} in (2.19).

We note in passing that the perfect $r_{ss'}$ (2.35) substituted into the weight (2.27-29) gives

$$w = \left(\frac{\psi_0(s(0))}{\psi_0(s(\tau))} \right) \cdot e^{-E_0\tau} = w_{transition} \cdot w_{diagonal} \, . \tag{2.37}$$

This implies that the unweighted distribution of arrival points $p(s,\tau)$ must approach $|\psi_0(s)|^2$ in the perfect guidance case as τ goes to infinity, just as in the continuum algorithm. This follows because the weight (2.37) is itself proportional to $1/\psi_0(s)$, and a histogram of these weights by construction approaches $k\psi_0(s)$ at large τ.

Application to the Heisenberg antiferromagnet

As a trial application of the algorithm we carried out a simulation of the one dimensional Heisenberg antiferromagnet [24], which is described by the Hamiltonian

$$H_{1d\ Heisenberg} = \sum_{i=1}^{N} \vec{S}_i \cdot \vec{S}_{i+1} = H_0 + H_I \, , \tag{2.38}$$

$$H_0 = \sum_{i=1}^{N} S_i^z \, S_{i+1}^z \equiv \sum_s V_0(s) \, |s\rangle\langle s| \, , \tag{2.39}$$

$$H_I = -\frac{1}{2} \sum_{i=1}^{N} (S_i^+ S_{i+1}^- + S_i^- S_{i+1}^+) \; . \tag{2.40}$$

The minus sign in H_I which is required for positive transition weights is induced by a change of phase of odd-numbered $| \downarrow \rangle$ states.

To simulate the ground state of this system we started random walks from a Néel configuration at $\tau = 0$ and followed their evolution in Euclidean time. The guidance matrix used was of the form

$$r_{ss'} = R - S \cdot (V_0(s') - V_0(s)) \; , \tag{2.41}$$

with R and S adjustable constants. ($r_{ss'}$ was not chosen to be of the ideal form (2.36) because that result had not yet been derived.) The guiding constants R and S were estimated by minimizing the variance of the log of the total weight factor (2.27) numerically. Systems having up to 48 spins were studied, and the measured energy per spin on $N = 48$ was

$$\left. \frac{E_0}{N} \right|_{N=48} = -0.4434 \pm 0.0004 \; , \tag{2.42}$$

which compares well with the expected value of -0.4435. This is the longest Heisenberg chain simulated at $T = 0$ in a reported Monte Carlo calculation to date, and although this measurement required the generation of 2^{20} random walks, the CPU time used was an encouragingly modest 33 microVAX-II hours. Measured values of E_0/N versus N clearly showed an approach to a value consistent with the exact result of Bethe [26],

$$\lim_{N \to \infty} \frac{E_0}{N} = \frac{1}{4} - \ln 2 \approx -0.4431 \; . \tag{2.43}$$

We also studied the triplet excited state energies, which were generated by starting each random walk from a Néel state with a single flipped spin. With increasing N the triplet-singlet gap $E_1 - E_0$ decreased approximately as k/N, but on the longest chains studied ($N = 40$ and 48) a bias of $\delta(E_1 - E_0) \approx .05$ appeared. High statistics runs carried out recently on the Caltech hypercube by Kotchan [27] suggest that the discrepancy was due to insufficient statistics in the triplet (E_1) measurement.

Now we will describe an application of the random walk algorithm to the two dimensional Heisenberg antiferromagnet. This system is described by the Hamiltonian

$$H = H_0 + H_I \; , \tag{2.44}$$

$$H_0 = \sum_{<ij>} S_i^z S_j^z \; , \tag{2.45}$$

$$H_I = -\frac{g}{2} \sum_{<ij>} (S_i^+ S_j^- + S_i^- S_j^+) . \tag{2.46}$$

with $g = 1$ being the isotropic case. Quantities of particular interest in this model are the ground-state energy per spin E_0/N^2, the triplet-singlet energy gap $E_1 - E_0$ and the possibility of long range antiferromagnetic order in the ground state. The latter is measured through the squared "staggered magnetization", defined by

$$\mathbf{N}_z^2 \equiv \langle\psi_0| \left(\frac{2}{N^2} \sum_i (-)^{(i_x+i_y)} S_i^z \right)^2 |\psi_0\rangle . \tag{2.47}$$

Various normalizations for this quantity are currently in use; the choice (2.47) is convenient because it gives $\mathbf{N}_z^2 = 1$ for a Néel state, which has the maximum possible antiferromagnetic alignment.

For this application we have chosen a guiding matrix $r_{ss'}$ which is an exponential in the change of diagonal energy V_0, and which is also of the form (2.36),

$$r_{ss'} = \frac{g}{2} \exp\Big(- c\, \Delta V_0 \Big) . \tag{2.48}$$

This form encourages the random walks to step preferentially towards configurations having lower "classical" energy V_0 for $c > 0$. Details of this application have been discussed elsewhere [25], so here we simply summarize our results for singlet and triplet energies on $N \times N$ square lattices, which are

$$\lim_{N\to\infty} \frac{E_0}{N^2} = -0.672 \pm 0.001 \tag{2.49}$$

and

$$E_1 - E_0 \approx \frac{2.1}{N} , \tag{2.50}$$

$$\lim_{N\to\infty} E_1 - E_0 = 0.052 \pm 0.085 \approx 0. \tag{2.51}$$

These numbers were obtained by extrapolation from Monte Carlo measurements on 4×4, 6×6 and 8×8 lattices. (We unnecessarily used the total weight (2.27) because we had not yet proven that the transition weight could be discarded for energy measurements, given an $r_{ss'}$ of the form (2.36).)

Previous results for E_0/N^2 quoted in the literature vary between -0.641 [28] for a variational calculation and -0.716 [29] for a perturbative technique, and a numerical extrapolation due to Oitmaa and Betts using results from lattices up to 4×4 in extent gave -0.655 ± 0.005 [30]. These and other earlier results are reviewed in [25]; numbers between -0.64 and -0.67 have often been cited. Recently the intense interest in two dimensional spin systems has stimulated studies which have found values close to -0.670. Specifically, the numbers and the methods employed are -0.668 ± 0.002 [31] (extrapolated Ising model perturbation theory); -0.672 [32] (real space renormalization group); -0.670 ± 0.002 [33] (finite temperature Monte Carlo); -0.672 ± 0.001 [34] (direct diagonalization on up to 26-site lattices); -0.668 [35] (RVB variational); and -0.675 ± 0.010 [36] (direct diagonalization on up to 24-site lattices). Our energy measurements and the direct diagonalization results are shown in figure 2, which is taken from Tang and Hirsch [34]. (Their convention gives energies scaled up by a factor of two.) We note in passing that the discrepancy in E_0/N^2 in figure 2 does not appear in our more recent unpublished results on the 6×6 lattice. Our previous τ extrapolation apparently introduced a significant systematic error. The consensus of the recent work is that the ground state energy of the two dimensional isotropic Heisenberg antiferromagnet is approximately

$$\lim_{N\to\infty} \frac{E_0}{N^2} = -0.670 \pm 0.002 \ . \tag{2.52}$$

Our finding that the triplet-singlet energy gap appears to be zero was previously conjectured [37-38] (see the discussion in [25]) and supporting evidence was seen in a finite temperature Monte Carlo calculation [39]. Tang and Hirsch [34] find triplet-singlet energy gaps on smaller lattices that also appear to approach zero as N goes to infinity, but they suggest convergence as k/N^2 rather than the k/N which our earlier results supported. It should be possible to resolve this issue using accurate energy measurements on moderately large lattices in future work.

There are several ways these quantities relate to the physics of high-T_c superconductors and their spin system models. Three examples are given below.

(1) E_0/N^2 may be used as a "benchmark" for comparison with variational calculations, to see which Ansätze include numerically important features in the ground state wavefunction.

(2) The energy gap $E_1 - E_0$ can be inferred experimentally through measurement of the spin-wave dispersion relation and possibly through the low temperature behavior of the specific heat.

(3) A suppression of electron magnetic moments on the copper ions relative to the free Cu^{++} value is observed in neutron scattering (see for example [40] and references cited therein). It has been suggested that this may be due to quantum fluctuations in the ground state which

suppress the measured moment relative to the free ion value, by an amount proportional to the staggered magnetization (2.47);

$$\frac{\mu(\mathrm{Cu}^{++})_{in\ situ}}{\mu(\mathrm{Cu}^{++})_{free\ ion}} = \sqrt{\mathbf{N}_z^2}\ . \tag{2.53}$$

The test suggested in (3) is actually controversial, as it compares the zero temperature magnetization in a two dimensional model with a finite temperature measurement in three dimensions. As the Mermin-Wagner theorem [41] says that the finite temperature staggered magnetization of the two dimensional isotropic antiferromagnet must be zero, it may be that a three dimensional model or an anisotropic coupling is actually required to describe this moment suppression.

Experimentally, the observed suppression is

$$\frac{\mu(\mathrm{Cu}^{++})_{in\ situ}}{\mu(\mathrm{Cu}^{++})_{free\ ion}} \approx \frac{0.6\,\mu_B}{1.8\,\mu_B} = 0.3\ . \tag{2.54}$$

This is indeed comparable to most published estimates of the ground state staggered magnetization of the two dimensional Heisenberg antiferromagnet. For example, Tang and Hirsch [34] find

$$\sqrt{\mathbf{N}_z^2} = \sqrt{0.0852 \pm 0.0181} \tag{2.55}$$

given our normalization convention. This appears to support the validity of a comparison between the two dimensional theory (2.53) and experiment (2.54). Caution is advisable however because variational results [35] indicate that states having moderate and zero staggered magnetization are very close in their expectation values of the energy. A slight perturbation on the Heisenberg model, such as interactions with the other spins in the crystal lattice, might significantly change the predicted magnetization (2.53) and invalidate the agreement with (2.54).

We have found that it is difficult to obtain accurate results for the staggered magnetization on lattices larger than 4×4; as is often the case, measurements of expectation values have much larger statistical errors than measurements of energies. We are currently attempting to extract $\mathbf{N}_z^2$ using the Feynman-Hellman theorem. This technique is of special interest because it relates the ground state expectation value of a quantity Q to the energy of a modified Hamiltonian;

$$\langle\psi_0|Q|\psi_0\rangle = \left(\frac{\partial}{\partial\lambda}E_0(\lambda)\right)\Big|_{\lambda=0} \tag{2.56}$$

where

$$E_0(\lambda)\,|\psi_0(\lambda)\rangle = (H + \lambda Q)\,|\psi_0(\lambda)\rangle\ . \tag{2.57}$$

In terms of random walk weight factors, (2.56) implies that expectation values are given by a large τ limit of weighted integrals of Q along each walk,

$$\langle\psi_0|Q|\psi_0\rangle =$$

$$\lim_{\tau_1,\tau_2\to\infty} \frac{1}{(\tau_2-\tau_1)}\left\{ \frac{< w(\tau_2)\int_0^{\tau_2} Q(s(\tau))d\tau >}{< w(\tau_2) >} - \frac{< w(\tau_1)\int_0^{\tau_1} Q(s(\tau))d\tau >}{< w(\tau_1) >} \right\} \tag{2.58}$$

As we can accurately measure energies using the random walk algorithm, one might anticipate that (2.56) would lead to accurate matrix elements. In practice this advantage is offset by the errors implicit in numerical differentiation. As a trial application of the Feynman-Hellman technique we have generated values for $\mathbf{N}_z^2$ on a 4×4 *anisotropic* Heisenberg antiferromagnet using (2.58); these are shown in figure 3. These results required a modest 4096 configurations at each of several τ values and have statistical errors of about ± 0.01. Unfortunately, larger lattices show slower convergence in τ and have larger statistical errors, so much longer runs will be required to study the long range order as a function of lattice size and model parameters. We hope to carry out measurements of $\mathbf{N}_z^2$ and other matrix elements of relevance to high-T_c superconductors using (2.58) in future work.

CONCLUSIONS

We have described a Hamiltonian Monte Carlo algorithm that can be used to obtain estimates for energies and matrix elements of low lying states of a specified quantum Hamiltonian system. The technique relies on an isomorphism between the Schrödinger equation in Euclidean time and a classical diffusion problem to generate quantum expectation values using random walks in configuration space. The algorithm has two variants which are applicable to continuous and discrete Hilbert spaces. We have shown that an optimum importance sampling exists in walk generation which allows energy measurements with zero error. An application of the discrete algorithm to the Heisenberg antiferromagnet was reviewed; this system is of current interest as a model of the high temperature superconductors. We obtained an accurate estimate of the ground state energy of this system which agrees with other recent work. The status of our attempts to measure the staggered magnetization (a crucial order parameter for high-T_c superconductivity) was also discussed. This staggered magnetization measurement and the optimization of the discrete algorithm summarize work carried out immediately after the meeting and thus constitute a "note added in proof".

ACKNOWLEDGEMENTS

It is a pleasure to thank the organisers of "Computational Atomic and Nuclear Physics at One Gigaflop", in particular C.Bottcher, M.R.Strayer and J.B.McGrory, for their kind invitation to the meeting and for the opportunity to present this material. I would also like to thank my colleagues G.J.Daniell and D.Storey (at the University of Southampton) and D.Kotchan and E.S.Swanson (at the University of Toronto) for the pleasure of collaborating with them on various aspects of this work. The applications described in this paper have required large amounts of CPU time, and I am especially grateful to G.C.Fox, A.Jacobs and A.Griffin, the Caltech Hypercube group and the Ontario Centre for Large Scale Computation for making available adequate computing resources. The financial support of the Natural Sciences and Engineering and Research Council of Canada, the Ontario Centre for Materials Research and the Universities of Guelph and Toronto is gratefully acknowledged.

REFERENCES

[1] J.G.Bednorz and K.A.Müller, Z. Phys. B**64** 189 (1986).

[2] J.Bardeen, L.N.Cooper and J.R.Schrieffer, Phys. Rev. **108** 1175 (1957).

[3] Novel Superconductivity, S.A.Wolf and V.Z.Kreskin editors (Plenum, 1987).

[4] R.J.Cava, B.Batlogg, J.J.Krajewski, R.Farrow, L.W.Rupp Jr, A.E.White, K.Short, W.F.Peck and T.Kometani, Nature **332** 814 (1988).

[5] T.M.Rice, Nature **332** 780 (1988).

[6] J.M.Longo and P.M.Raccah, J. Sol. State Chem. **6** 526 (1973).

[7] Y.Yamaguchi, H.Yamauchi, M.Ohashi, H.Yamamoto, N.Shimoda, M.Kikuchi and Y.Syono, Jpn. J. Appl. Phys. Part **2** L447 (1987).

[8] M.W.Shafer, T.Penney and B.L.Olson, Phys. Rev. D**36** 4047 (1987).

[9] P.W.Anderson, *Science* **235** 1196 (1987).

[10] P.W.Anderson, G.Baskaran, Z.Zou and T.Hsu, Phys. Rev. Lett. **58** 2790 (1987).

[11] S.Kivelson, P.S.Rokhsar and J.M.Sethra, Phys. Rev. B**35** 8865 (1987).

[12] J.E.Hirsch, Phys. Rev. Lett. **59** 228 (1987).

[13] F.C.Zhang, "High T_c Superconductivity - Hubbard Model?", ETH-Hönggerberg preprint (1987).

[14] K.Huang and E.Manousakis, "Strong-Coupling Hubbard Model of High Temperature Superconductivity", MIT preprint CTP-1490 (July 1987).

[15] J.R.Schrieffer, X.-G.Wen and S.-C.Zhang, Phys. Rev. Lett. **60** 944 (1988).

[16] J.E.Hirsch, S.Tang, E.Loh Jr and D.J.Scalapino, "Pairing Interaction in Two-Dimensional CuO_2".

[17] S.H.Liu, "The Itinerant Resonating-Valence-Bond Model for Superconductivity", Oak Ridge National Laboratory preprint (August 1987).

[18] M.H.Kalos, Phys. Rev. **128** 1791 (1962).

[19] J.E.Hirsch, R.L.Sugar, D.J.Scalapino and R.Blankenbecler, Phys. Rev. B**26** 5033 (1982).

[20] M.H.Kalos and P.Whitlock, Monte Carlo Methods (Wiley 1986).

[21] T.Barnes, G.J.Daniell and D.Story, Nucl. Phys. **B265** [FS15] 253 (1986).

[22] T.Barnes and D.Kotchan, Phys. Rev. **D35** 1947 (1987).

[23] T.Barnes, Toronto preprint UTPT-87-16.

[24] T.Barnes and G.J.Daniell, Phys. Rev. **B37** 3637 (1988).

[25] T.Barnes and E.S.Swanson, Toronto preprint UTPT-88-01, Phys. Rev. **B37** 9405 (1988).

[26] H.Bethe, Z. Phys. **71** 205 (1931).

[27] D.Kotchan, unpublished.

[28] H.Taketa and T.Nakamura, J. Phys. Soc. Japan **11** 919 (1956).

[29] M.H.Boon, N. Cimento **21** 885 (1961).

[30] J.Oitmaa and D.D.Betts, Can. J. Phys. **56** 897 (1978).

[31] D.A.Huse, Phys. Rev. **B37** 2380 (1988).

[32] D.C.Mattis and C.Y.Pan, "Ground State Energy of Heisenberg Antiferromagnet for Spins S=1/2 and S=1 in d=1 and 2 Dimensions.".

[33] J.D.Reger and A.P.Young, Phys. Rev. **B37** 5978 (1988)

[34] S.Tang and J.E.Hirsch, "Long Range Order without Broken Symmetry - 2D Heisenberg Antiferromagnet at Zero Temperature." (27 February 1988).

[35] S.Liang, B.Doucot and P.W.Anderson, "Some New Variational RVB-type Wave Functions for the Spin 1/2 Antiferromagnetic Heisenberg Model on a Square Lattice." (29 February 1988).

[36] E.Dagotto and A.Moreo, University of Illinois preprint ILL-(TH)-88-13 (April 1988), Phys. Rev. B (to appear).

[37] E.Lieb, T.Schultz and D.Mattis, Ann. Phys. (N.Y.) **16** 407 (1961).

[38] I.Affleck, Phys. Rev. **B37** 5186 (1988).

[39] D.H.Lee, J.D.Joannopoulos and J.W.Negele, Phys. Rev. **B30** 1599 (1984).

[40] W-H.Li *et al* , "Long Range Order of the Cu in Oxygen Deficient $R\mathrm{Ba_2Cu_3O_{6+x}}$.".

[41] N.D.Mermin and H.Wagner, Phys. Rev. Lett. **17** 1133 (1966).

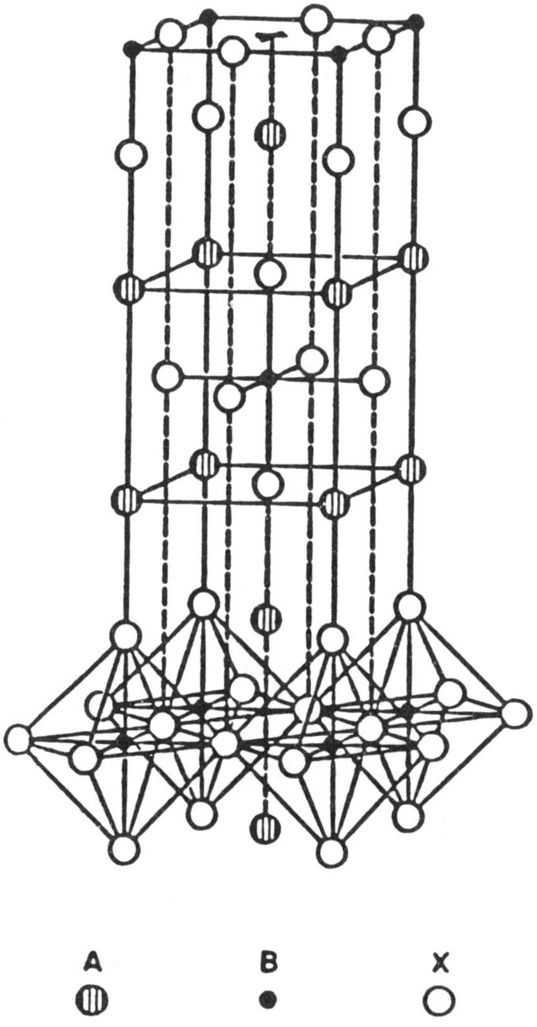

Figure 1. The unit cell of La_2CuO_4 [4]. (A=La, B=Cu and X=O.)

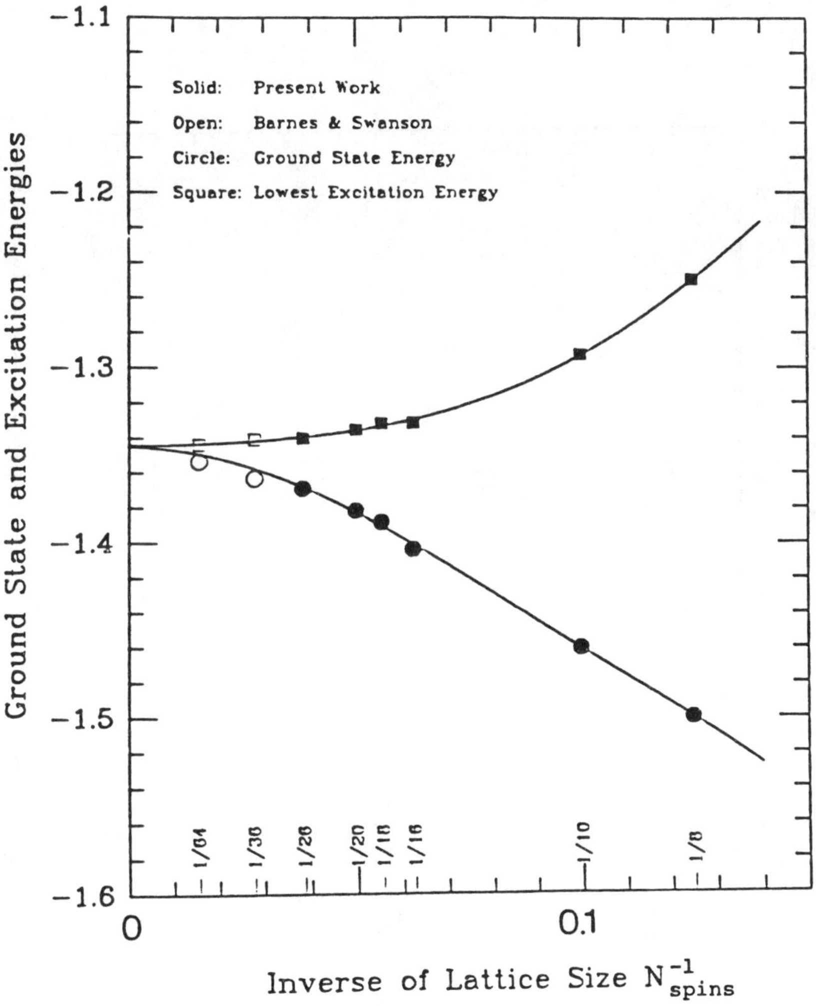

Figure 2. Singlet and triplet 2DHA energies (Tang and Hirsch [31]).

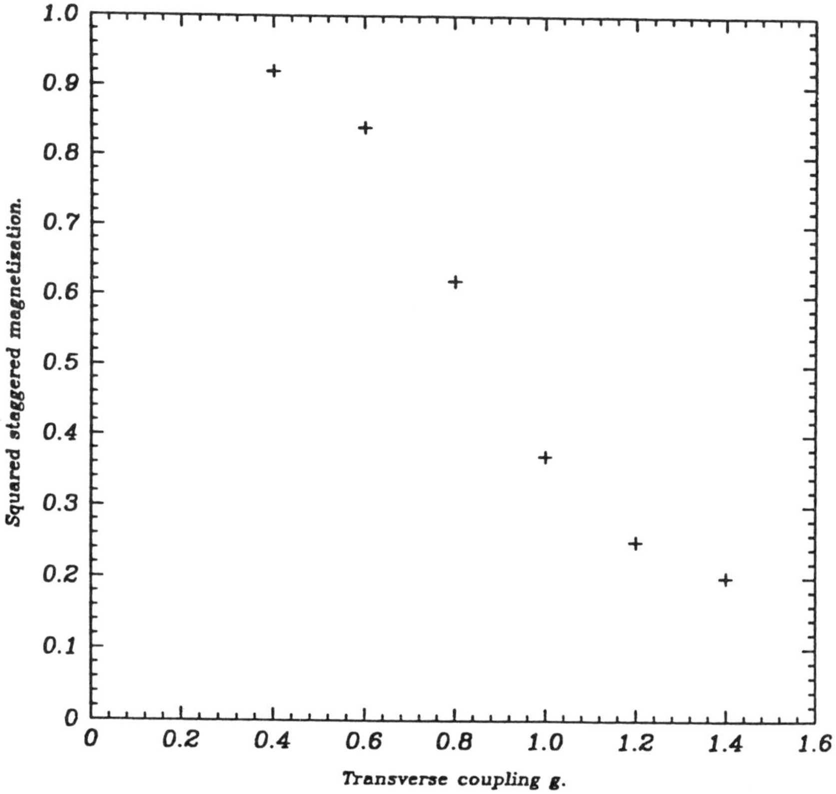

Figure 3. N_z^2 magnetization of the 4×4 anisotropic 2DHA.

MANY-BODY EFFECTS IN ION-ATOM COLLISIONS

R. E. OLSON
Physics Department, University of Missouri-Rolla,
Rolla, Missouri 65401, U.S.A.

Abstract The advent of vector- and parallel-processing computers is making it possible to incorporate the dynamics of all interacting particles (including electrons) in scattering calculations. Classical simulations of intermediate energy ($E \approx 10$ MeV/u) collisions of multiply-charged ions with many electron target atoms are found to be able to predict and reproduce experimental data with good accuracy. Further extensions of theory will make it possible, for the first time, to calculate energy deposition by ions in solids and biological material. Such extensions have direct relevance to heavy ion cancer therapy and cosmic ray induced semiconductor upsets in satellites and space stations.

INTRODUCTION

The advent of supercomputers is making it possible to address an interesting, and important, set of intermediate energy (1 - 20 MeV/u) ion-atom collision problems. These collisions are characterized by highly-charged ions such as Fe^{22+} or U^{65+} impinging on a many-electron target atom such as C or Ar. The resulting collision exhibits many-body phenomena in that all the electrons and the two nuclei participate in the scattering reaction. The reaction is complex, and typically results in the emission of several electrons from the collision region along with the capture of one or more electrons by the projectile ion:

$$A^{q+} + B \rightarrow A^{(q-i)+} + B^{j+} + (j-i)e. \qquad (1)$$

The description of these scattering events is a true, n-body problem which requires appropriate approximations and sufficient computer time. Only recently has it been possible to numerically follow the time evolution of the nuclei, and electrons, during such collisions. Successful benchmarking of the calculations against atomic collision data will allow future extensions to provide a basic understanding of ion implantation mechanisms, heavy ion cancer therapy, and cosmic ray induced semiconductor upsets in satellites.

THEORETICAL METHOD

The theoretical method we have employed to determine the dynamics of MeV/u collisions of highly-stripped ions with atoms (Reaction 1) is based on the numerical solution of the classical Hamiltonian for all particles participating in the collision. The Hamiltonian is written

$$H = \frac{p_a^2}{2m_a} + \frac{p_b^2}{2m_b} + \sum_{i=1}^{N} \frac{p_i^2}{2m_e} + \sum_{i=1}^{N} \left(\frac{z_a z_i}{R_{ai}} + \frac{z_b z_i}{R_{bi}} \right) + \frac{z_a z_b}{R_{ab}} , \qquad (2)$$

where N is the number of electrons on target atom B, and the indices a and b represent the projectile ion and target nucleus, respectively. From Eq. (2), one obtains a set of 6(N + 2) coupled, first-order differential equations arising from the necessity to determine the time evolution of the xyz Cartesian coordinates of each particle

$$\frac{dc_j}{dt} = \frac{\partial H}{\partial p_j} , \qquad (3)$$

and their xyz momenta

$$\frac{\partial p_j}{\partial t} = - \frac{\partial H}{\partial c_j} .$$

The electrons are initially placed in a microcanonical distribution of orbits about the target nucleus with an effective interaction charge determined by the sequential binding energies of the target atom. The 6(N + 2) coupled equations are then numerically solved for a Monte Carlo determined distribution of electron orbits and impact parameters for the collision.

In practice, we routinely need about 5,000 different trajectories to obtain the total cross sections for Reaction (1). Cross sections differential in angle or energy require several tens of thousand trajectories. For a case such as U^{55+} + Ar where we consider the uranium ion to be approximated by a point charge, but include the K, L, and M-shells of the argon atom, 120 coupled equations are numerically integrated. The nominal amount of CRAY-1 CPU time required for 10,000 trajectories is approximately 5 hours. It is obvious such calculations were quite tedious before the advent of supercomputers.

NUMERICAL RESULTS

Illustrative examples of calculated cross sections for U^{q+} + Ar collisions at energies available to present accelerators are shown in Figs. 1-4. The calculated values are in good agreement with experimental observations[1] and nicely show the n-body nature of single electron capture which proceeds with the ionization of ~8 other electrons from the scattering center.

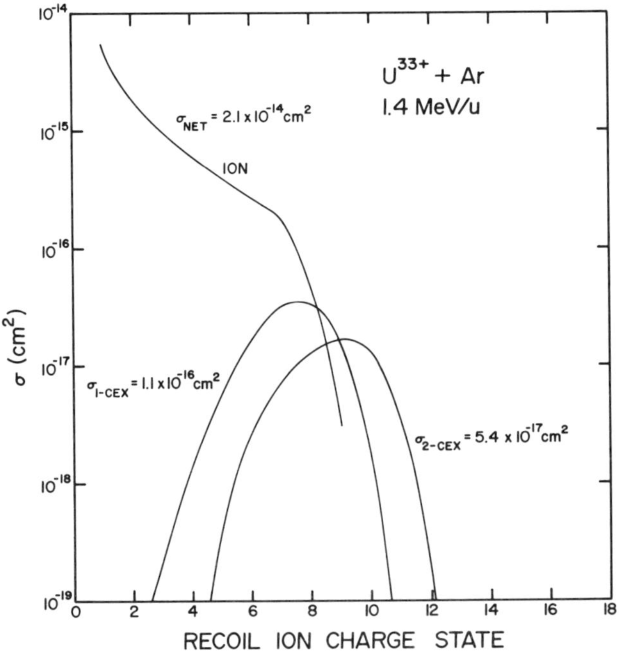

FIGURE 1 Multiple ionization and single and double electron capture cross sections for 1.4 MeV/u U^{33+} + Ar collisions.

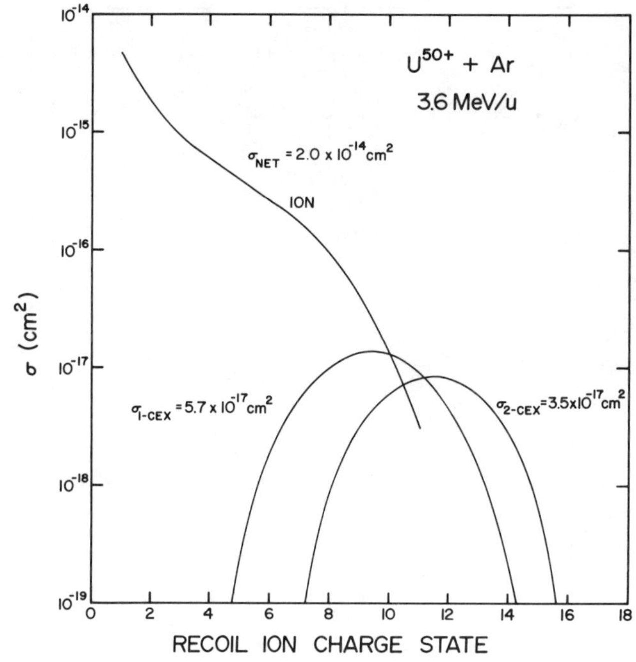

FIGURE 2 Multiple ionization and single and double electron capture cross sections for 3.6 MeV/u U^{50+} + Ar collisions.

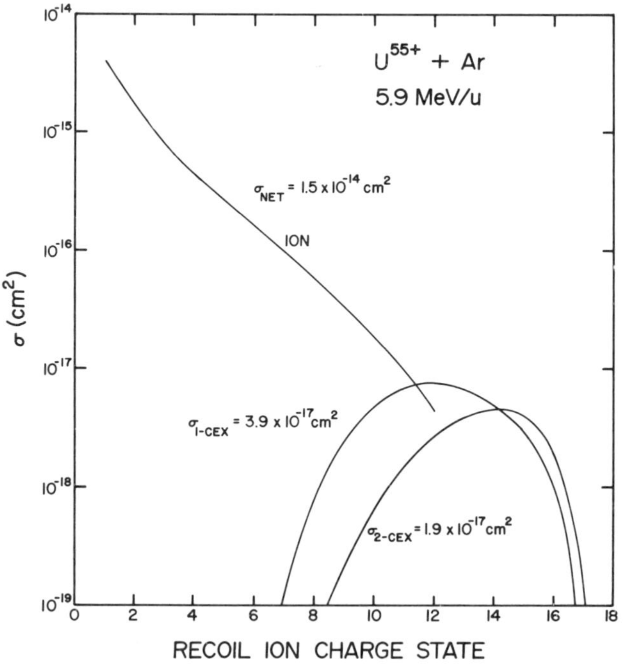

FIGURE 3 Multiple ionization and single and double electron capture cross sections for 5.9 MeV/u U^{55+} + Ar collisions.

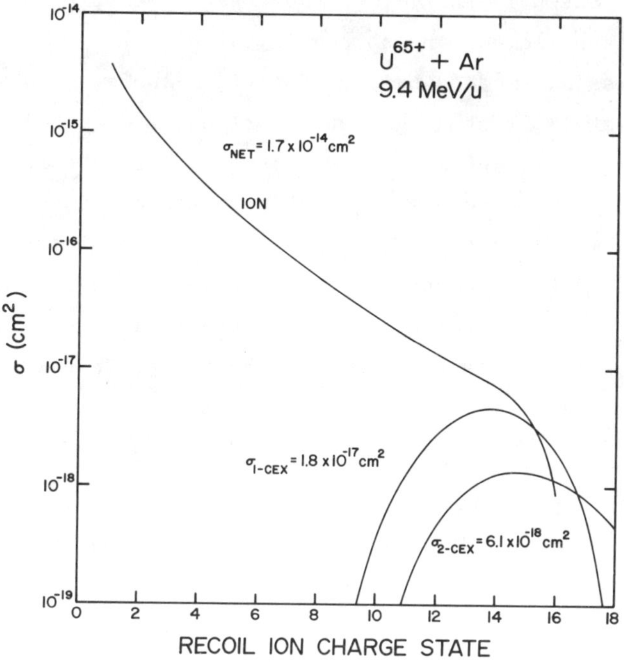

FIGURE 4 Multiple ionization and single and double electron capture cross sections for 9.4 MeV/u U^{65+} + Ar collisions.

It is interesting to note the large magnitude of the cross sections (on the order of 10^{-15} cm^2) for the production of low charge states of the recoiling target nucleus. Also illustrated is the decreasing importance of charge exchange with increasing projectile velocity. This is due to the increase mismatch in the velocity of the projectile ion with the L-shell electrons on the Ar target.

The classical method briefly described above has also been successfully tested against experimental data as to the recoil energies of the ionized target. It is surprising that of the ~GeV energy available in the collision, only approximately a few eV are given to the recoiling target nucleus.[2] Other comparisons to available experimental data on the angular scattering of the projectile[3] and the energy and angular distributions of the ionized electrons[4] indicate the validity of this numerical technique.

CONCLUDING REMARKS

The numerical method described above requires considerable CPU time, but requires little memory. In fact, the code easily runs on an IBM-PC. Hence, there exists the possibility to further extend the program by the use of a parallel processor. Such an extension can be readily made since individual trajectories do not require information from other events. Thus, a computer such as a HyperCube with 1024 processors will be ideal for this work and increase through-put by over an order-of-magnitude.

The availability of such a machine will allow one to place an array of atomic or molecular targets before the incident projectile and still obtain good statistics on the various collision events. In fact, the computer code is now being rewritten to determine the collision physics of the stopping and damage caused by energetic ions in silicon in order to provide the input necessary for design of computer chips for space applications.[5] Similar work on cell destruction by heavy ion beams is now providing a basis for an understanding of the mechanisms responsible for cell deactivation in heavy ion cancer therapy.[6] Considerable further work is needed, but the availability of gigaflop computing speeds will provide the opportunity to extend basic atomic collision codes to the transversal of ions through complex materials.

ACKNOWLEDGEMENT

This work was supported by the Office of Fusion Energy of the U. S. Department of Energy.

REFERENCES

1. H. Berg, R. Dörner, C. Kelbch, S. Kelbch, J. Ullrich, S. Hagmann, P. Richard, H. Schmidt-Böcking, A. S. Schlachter, M. Prior, H. J. Crawford, J. M. Engelage, I. Flores, D. Loyd, J. Petersen, and R. E. Olson, J. Phys. B (in press).
2. R. E. Olson, J. Ullrich, and H. Schmidt-Böcking, J. Phys. B 20, L809 (1987).
3. R. Schuch, H. Schöne, P. D. Miller, H. F. Krause, P. F. Dittner, S. Datz, and R. E. Olson, Phys. Rev. Lett. 60, 925 (1988).
4. C. Kelbch, R. E. Olson, S. Schmidt, H. Schmidt-Böcking, and S. Hagmann, J. Phys. B, (submitted).
5. D. Schultz (private communication).
6. S. Schmidt, J. Euler, C. Kelbch, S. Kelbch, R. Kock, G. Kraft, R. E. Olson, J. Ullrich, H. Schmidt-Böcking, Rad. Research (submitted).

ATOMIC STRUCTURE CALCULATIONS ON THE CRAY X-MP

CHARLOTTE FROESE FISCHER
Vanderbilt University, Box 6035B,
Nashville, TN 37235 USA

Abstract Atomic structure calculations require both radial and angular integrations, where the latter are often based on Racah algebra. With relatively minor modifications, good performance is obtained on vector machines for radial integrations. Angular integrations, however, present the bottleneck. Some recent improvements in the algorithms for angular integrations are described, as well as some multitasking experiments on the CRAY X-MP and CRAY 2. These show that the workload can easily be distributed evenly among available processors with dynamic scheduling.

INTRODUCTION

The many-body problem in quantum mechanics is well-known for its complexity. As Mort La Brecque[1] wrote, ''Once the number of bodies interacting with each other grows beyond two, the problems of computing their effects on each other begin to grow out of control.'' The many-body problem requires the solution of the equation,

$$H\Psi = E\Psi.$$

In this equation, H is a partial differential operator of dimension $3N$, where N is the number of electrons, Ψ is the total wave function describing a state of the system, and E is the total energy of the system.

For a one-electron atom, under the assumptions of a

nucleus with infinite mass and a point charge, the quantum mechanical problem can be solved exactly, both in non-relativistic and relativistic theory. With the addition of only one more electron, and with the same assumptions, the atomic structure problem requires the solution of a three dimensional partial differential equation. In the 1960's, Pekeris and his collaborators[2] developed expansion techniques for solving this non-relativistic problem which, even today, are used as a standard for comparison, but unfortunately these methods could not be extended readily to the ''many-body'' problem.

The multiconfiguration Hartree-Fock (MCHF) method[3], developed for bound state atomic system, usually performs well on systems that can be modeled as two-electron systems, but as the number of electrons increases the computational task rapidly becomes out of hand and simplifying assumptions are introduced that decrease the level of accuracy in the prediction of atomic properties.

The multiconfiguration Hartree-Fock and associated configuration interaction theories, including the relativistic multiconfiguration Dirac-Fock (MCDF) theory[4], are based on the variational principle and have been applied to a large number of systems. In fact, the single configuration Hartree-Fock results usually are taken as the starting point in many-body perturbation theory (MBPT). In MCHF theory,

$$\Psi \approx \sum_{1}^{m} c_i \Phi_i$$

where the configuration state function (CSF), Φ, is an antisymmetrized product of one-electron spin-orbitals (one for each electron) in which only the radial function is unknown. The radial functions are solutions of coupled systems of integro-differential equations that depend on the expansion coefficients, c_i. The latter are components of an eigenvector of an interaction matrix of order m, the entries of which depend on the radial functions. Thus the two problems are coupled and usually are solved iteratively.

The special case with $m = 1$, corresponds to an independent particle model, and so the error in the energy of the system predicted by this calculation is referred to as a ''correlation'' error. For simple systems such as the ground state of beryllium, the correlation error can be estimated. As m increases, an expansion exists[5] for which the first few configuration states capture a large percent of the correlation energy, but one rapidly reaches a region of diminishing returns where doubling the expansion length has only a small effect on the total energy. It is this ''accuracy barrier'' which makes the many-body problem in atoms such a challenge, at least in few electron systems. For larger systems, such as the transition metals with many electrons is unfilled shells, the complexity of the wave function expansion itself, which grows rapidly as correlation effects are included, presents a difficulty.

A number of general purpose codes are available for the prediction of atomic properties for bound state systems. These can be divided into two categories:

1. Non-relativistic with optional Breit-Pauli corrections.

2. Relativistic with Breit and Lamb shift corrections

A general purpose, atomic structure package based on the

MCHF[3] method falls in the former category whereas the MCDF[4] code belongs to the latter. One major difference in design of these two codes is the fact that MCHF is a series of programs, whereas MCDF has all steps of a calculation embodied in one big program.

In this paper we will outline the stages of an MCHF atomic structure calculation, describe the modifications that were required for improving performance on the CRAY X-MP, including some multitasking experiments, and show how these codes were used to predict a stable Ca^- ion[6], contrary to the general belief that the alkaline earths did not possess a positive electron affinity[7].

STAGES IN ATOMIC STRUCTURE CALCULATIONS

Atomic structure calculations generally have three distinct phases:

1. Generating lists of configuration state functions that define the wave function expansion
2. Performing angular integrations that express $\langle\Psi_f|Op|\Psi_i\rangle$ in terms of radial integrals
3. Performing radial integrations which include:
   - Solving differential equations that define radial functions
   - Performing definite or indefinite integrations

As an example, let us consider the computation of transition rates for forbidden (E2 and M1) transitions in the $3p^2$ isoelectronic sequence of Silicon . In the MCHF atomic structure package, the radial functions are determined from a non-relativistic (LS) energy expression, say for ^{3}P, whereas the wave function for an LSJ state may be determined in the Breit-Pauli (LSJ) approximation that includes the spin-orbit mixing of many terms, so this expansion may be much longer. Such a calculation could proceed as follows:

1. Generate CSF lists
   - List for ^3P LS expansion
   - List for LSJ Breit-Pauli expansion
2. Perform angular integrations
   - Determine energy expression for ^{3}P list
   - Determine energy expression for Breit-Pauli interaction matrix
   - Determine expression for the transition matrix element

For $Z = 14$ to 24

3. Determine radial functions
4. Evaluate the Breit-Pauli interaction matrix
   - Compute eigenvalues and eigenvectors
   - Print energy levels
5. Evaluate the transition matrix elements
   - Compute transition rates
   - Print line list

In many problems, the same configuration list can be used for a whole isoelectronic sequence. Since angular integrations are independent of the atomic number, these need be performed only once whereas the remaining calculations are repeated for each value of Z.

In large atomic structure calculations, configuration lists are generated through the application of rules. This is not a CPU intensive phase and will not be considered further here. Let us consider the remaining phases separately since they have quite different characteristics.

RADIAL INTEGRATIONS

Some codes[8] express the radial functions in terms of an analytic basis of Slater type orbitals, $r^{\alpha}e^{-\beta r}$. Others[4,9], as well as MCHF, represent the radial functions as a vector of values. Such approaches benefit from the vectorization capabilities of supercomputers such as the CRAY X-MP.

The Hartree-Fock program has been benchmarked on a variety of computers. Some of the timing results are presented in Table I along with a relative speed compared with a VAX 11/780. It should be remembered that such timing studies compare the hardware as well as the compiler. This is particularly evident in the results included for the SUN 3/160 where full optimization greatly improves the performance over partial optimization.

The changes introduced into the code to improve vectorization were relatively modest. Profiling of the code had shown that, for large atoms, about 60% of the

TABLE I Hartree-Fock Timing studies for Radium $7s^2\,{}^1S$

	CPU Time (sec.)	Relative Speed
IBM PC	3354.	0.076
VAX 11/780	254.	1.000
SUN 3/160 with floating point board		
-p option	195.9	1.30
-O option	71.3	3.56
FPS 164	18.1	14.0
CYBER 205	25.5	10.0
CRAY X-MP		
Unmodified Code	3.25	78.
Modified Code	1.96	130.

CPU time was spent evaluating the integrals

$$Y^k(i,i';r) = \int_0^r (t/r)^k P_i(t)P_{i'}(t)dt + \int_r^\infty (r/t)^{k+1} P_i(t)P_{i'}(t)dt.$$

The implementation of these integrations in FORTRAN code involved the following recurrence:

```
DO I = 2, N
   X(I) = X(I-1) + (expression independent of X)
CONTINUE
```

Such recurrences inhibit vectorization. The performance was improved by replacing DO loops like the above by a pair of loops:

```
DO I = 2, N
   F(I) = (expression)
CONTINUE

DO I = 2, N
   X(I) = X(I-1) + F(I)
CONTINUE
```

With a few minor modifications like the above, a speedup of a factor of more than 100 was obtained over the VAX 11/780. No doubt, further improvements are possible, but angular integrations soon proved to be the bottle neck for large calculations on the CRAY X-MP.

ANGULAR INTEGRATIONS

Many atomic structure packages[3,4,8,10], for bound state and scattering properties alike, use Racah algebra for calculating the angular integrals, where each integral is expressed in terms of recoupling coefficients, and many have used the NJSYM program by Burke[11] to calculate the recoupling coefficients. Fraga et al.[9] expand the wave function in terms of determinants, so that simple rules based on permutations can be used instead, with the consequence that on a CRAY X-MP 75% of the CPU time is consumed by a subroutine PERMUT that permutes the orbitals.

Some preliminary investigations for the energy levels of U^{+4}, using the MCDF code, dramatically illustrate the extent of the problem. An average energy level calculation for U^{+4} $5d^{10}5f^{2}6s^{2}6p^{6}$ very quickly produces results for which there are a number of energy level inversions when compared with experimental data. First order perturbation theory[12] has suggested that the most important correlation effects omitted in this calculation arise from the configuration $5d^8 5f^4 6s^2 5p^6$. A wave function expansion for only the $J = 0$ states requires 123 configuration state functions (representing the different possible couplings). It was estimated that 13 hours of CRAY X-MP time would be required to perform the angular integrations that define the 123×123 interaction matrix. For $J = 6$, the interaction matrix was more than 300×300 and for the intermediate J-values the size was expected to be even larger. Correlation studies in such complex atoms are not feasible at this time, not only because of CPU time, but also because of memory requirements. The MCDF code with dimensions set to 400×400 exceeded the allowed maximum program size on the CRAY X-MP.

A number of researchers have recognized the need to reduce the CPU requirements for angular integrations. These calculations are not vector in nature, and vector machines do not provide the speedup often associated with large calculations. Thus most of the improvements have resulted from refinements in the algorithms. Sott and Hibbert[13] analyzed the WEIGHTS program that performs the angular integrations for electrostatic interactions. They introduced a number of modifications that reduced the time by about a factor three. One of the changes involved a new routine for computing the Racah coefficients using a table of log($n!$). In the modified program, their test example still spent about 25% of the CPU time in this routine. Recently, Bar-Shalom and Klapisch[14] have published a recoupling package that uses graphical representations of recoupling, allowing expression simplification and also the use of ''free'' variables. In this way, successive calls with different values of the free variables, do not require a full re-evaluation of the expression, but rather a substitution process. We have found that their NJGRAF produces a further reduction in CPU time of about a factor two in WEIGHTS and a factor of three to four in the integration of Breit-Pauli operators. These modifications would increase the fraction of time spent evaluating Racah coefficients.

Recently, Scott et al.[15] published an implementation of the Racah routine in the programming language, Occam, using transputers in parallel. They reported on studies using both a 6 transputer network and a 4 transputer network. For their test cases, the speedups ranged from 2.4 to 5.1, for the former, and 2.4 to 3.1 for the latter.

A simpler approach is to use table lookup for the commonly used values. An analysis of a Breit-Pauli calculation for all possible couplings and terms of $3s^2 3p$, $3p^3$, $3s3p3d$, $3p3d^2$ with an allowed $J = 1/2$ or $J = 3/2$, indicated that 2% of all the calls (ignoring symmetry) had the same arguments. The RACAH routine has six integer arguments. If none is too big and the function value has already been defined, it is retrieved from a table; otherwise it is computed and saved in the table for future use. If any argument is too big, the value is computed. We have a trade-off situation here in that, with more memory, less values will need to be recomputed. One advantage of breaking a calculation into a sequence of programs (unlike MCDF, for example) is that

TABLE II CPU time in seconds of a Breit-Pauli calculation using different versions of the program on different computers

	SUN 3/160	VAX 11/780	CRAY X-MP
NJSYM	206.6	322.4	
NJGRAF	76.2		4.91
+ TRACAH	48.1		

not all the arrays need be in memory at the same time and more memory is available for a given program without relying on overlaying.

Table II shows some CPU times (in seconds) for the Breit-Pauli calculation mentioned earlier. The NJSYM version includes the modifications made by Scott and Hibbert, NJGRAF contains the new recoupling package, whereas NJGRAF + TRACAH contains the Racah routine with table look-up. The benchmarking is not nearly as complete, but it is clear that the speedup in going to the CRAY X-MP is not nearly as great as for radial integrations

In collaboration with N.S. Scott and Jungsoon Yoo[16], multitasking was evaluated as a means of reducing wall clock time. In fact, the problem of performing angular integrations lends itself readily to parallel execution at a fairly high level of granularity. The driver of the Breit-Pauli program and breit subroutine called by it have the following relatively simple structure:

```
PROGRAM driver              SUBROUTINE breit(i,j)
COMMON /data/...            COMMON /data/...
(initialization)            COMMON /work/...
DO i = 1,n
  DO j = 1,i                 (Many levels
    call breit(i,j)             of calls)
CONTINUE                     END
(postprocessing)
```

For multitasking, COMMON /work/ must become TASK COMMON /work/ so that each task has its own workspace. COMMON /data/ is shared by all tasks. Static load balancing was tested on the CRAY X-MP where the parent task created one child task, and each task performed half the calls to breit. In a Breit-Pauli calculation, where many calls may quickly determine a zero interaction, such a strategy may lead to an unbalanced work load. Dynamic load balancing was tried on the CRAY X-MP (with two processors) and also on the CRAY 2 (with 2 and 4 processors). In this implementation, each task performs the calculation for the next (i,j) pair until all calls are completed.

Figure 1. summarizes the results of this study for a case where the interaction matrix is sparse. Two types of times are depicted --- CPU time in seconds and real or wallclock time in minutes. Unitasking refers to the original program using only one CPU. The figure clearly shows that

1. CRAY 2 requires more CPU time than the CRAY X-MP

2. Multitasking incurs overhead

3. With multitasking the wallclock time is reduced

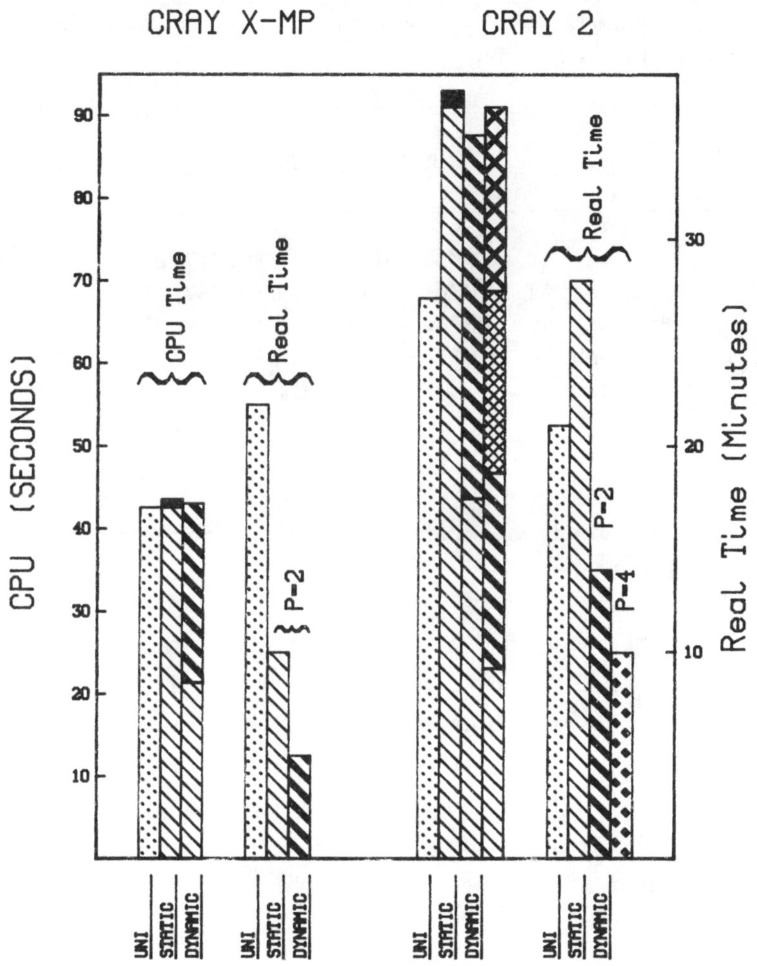

FIGURE 1. CPU and real time for some multitasking studies on the CRAY X-MP and the CRAY 2, with P processors. 1) UNI -- no multitasking 2) STATIC -- each processor performed half the cases 3) DYNAMIC -- each processor computed the next case.

TABLE III Total energies (in au) for MCHF calculations and predicted EA's for three different expansions for the ground state of Ca and the $4s^2 4p\ ^2P$ state of Ca^-

electrons	Ca $4s^2\ ^1S$	Ca^- $4s^2 4p\ ^2P$	EA (eV)
4s, 4p, 3d, 4f	−676.786181	−676.777788	−0.224
+ 5s, 5p, 4d	−676.787048	−676.787485	0.012
+ 6s, 6p, 5d, 5f	−676.787094	−676.789138	0.056
+ rel. shift	−679.566766	−679.568426	0.045
Experiment[17]			0.043±0.007

4. Dynamic load balancing performs better than static load balancing

It should be mentioned that these times were obtained in a multiprogramming environment and that the real time depends on the system load and would not be exactly reproducible. These studies suggest that further improvements in the speedup of angular integrations could be achieved easily through the use of system with a larger number of processors.

APPLICATION TO THE STUDY OF Ca^- AND Sr^-

The MCHF atomic structure package was used recently in the prediction of a stable $4s^2 4p\ ^2P$ state for Ca^- and Sr^-. The binding energy derived from the former[6] was in good agreement with the experimental value obtained by Pegg et al.[17] In these calculations, a rule based program was used to generate the configuration lists. One of the options, appropriate for small systems (in the present case the neutral atom was modeled as a core + two electrons, the negative ion as core + three electrons) is to generate all possible configurations and their couplings that can be formed from a given set of electrons. The same set was used for both the atom and the ion. As the set of possible electrons was increased, the length of the wave function expansion for ion grew quite rapidly to $m = 325$. Calculations for the two systems were performed independently, that is, the orbitals for the neutral atom and negative ion were not required to be the same. In the first calculation, all orbitals were varied, but in subsequent ones, only the new ones were determined variationally. The results for Ca and Ca^- are reported in Table III. As a final correction, relativistic shift corrections were included in the calculations. These had the effect of reducing the electron affinity slightly. Similar calculations for Sr^- predict a somewhat larger electron affinity[18].

This research was supported by the U.S. Department of Energy, Office of Basic Energy Sciences.

REFERENCES

1. M. La Brecque, MOSAIC **17**, 2 (1986).
2. C.L. Pekeris, B. Schiff, and H. Lifson, Phys. Rev. **126**, 1057 (1962).
3. C. Froese Fischer, Comput. Phys. Commun. **1**, 151 (1969): also Comput. Phys. Reports **3**, 274 (1986).
4. I.P. Grant, B.J. McKenzie, P.H. Norrington, D.F. Mayers, and N.C. Pyper, Comput. Phys. Commun. **21**, 207 (1980).

5. C. Froese Fischer, Int. J. Quant. Chem. **8**, 5 (1974).

6. C. Froese Fischer, J.B. Lagowski, and S.H. Vosko, Phys. Rev. Letters **59**, 2263 (1987).

7. H. Hotop and W.C. Lineberger, J. Phys. Chem. Ref. Data **14**, 731 (1985).

8. A. Hibbert, Computer Phys. Commun. **10**, 436 (1976).

9. S. Fraga, M. Klobukowski, J. Muszynska, K.M.S. Saxena, J.A. Sordo, J.D. Climenhaga and P. Clark, Comput. Phys. Commun. **47**, 159 (1987).

10. K.A. Berrington, P.G. Burke, J.J. Chang, A.T. Chivers, W.D. Robb, and K.T. Taylor, Comput. Phys. Commun. **8**, 149 (1974).

11. P.G. Burke, Comput. Phys. Commun. **1**, 241 (1970).

12. J. Morrison, K. Rajnak, and M. Wilson, J. de Phys. Colloque C4, **31**, C4-167 (1970).

13. N.S. Scott and A. Hibbert, Comp. Phys. Commun. **28**, 189 (1982).

14. A. Bar-Shalom and M. Klapisch (submitted to Comput. Phys. Commun.) (1987).

15. N.S. Scott, P. Milligan, and H.W.C. Ridley,Comp. Phys. Commun. **46**, 83 (1987).

16. N.S. Scott, C. Froese Fischer, J. Yoo, Parallel Computing (accepted) (1988).

17. D.J. Pegg, J.S. Thompson, R.N. Compton, and G.D. Alton, Phys. Rev. Letters **59**, 2267 (1987).

18. C. Froese Fischer, (in preparation) (1988).

HADRONIC STRUCTURE*

J. W. NEGELE

Center for Theoretical Physics, Laboratory for Nuclear Science, and Department of Physics, Massachusetts Institute of Technology. Cambridge, Massachusetts 02139 U.S.A.

ABSTRACT Two exploratory calculations demonstrate the promise of numerical calculations for fundamental study of the structure and interaction of hadrons. Exact solution of a confining quark model and comparison to an effective hadronic approximation to it provide an instructive test of the convolution approximation commonly used in describing quark degrees of freedom in nuclei. Stochastic solution of lattice QCD in two dimensions is shown to yield meson Bethe–Salpeter wavefunctions and density-density correlation functions which agree in detail with continuum weak coupling solutions. This success motivates analogous realistic calculations in four space-time dimensions.

I. INTRODUCTION

One of the primary challenges in contemporary nuclear physics is to understand the structure and interaction of hadrons in terms of quarks and gluons, and how this structure manifests itself in observed nuclear phenomena.

To understand low energy nuclear physics, we seek to extract from the full theory an effective theory in terms of hadron degrees of freedom alone, and to identify the domain of validity of this hadronic approximation. We need to identify and unambiguously interpret signatures of subnuclear degrees of freedom in nuclei such as the so-called EMC effect in deep inelastic lepton scattering.

* This work is supported in part by funds provided by the U. S. Department of Energy (D.O.E.) under contract #DE-AC02-76ER03069.

Current hadronic models are at best caricatures, emphasizing some particular aspect of QCD while neglecting others. For example, the bag model focuses on confinement, chiral models emphasize chiral symmetry, and the Skyrme model emphasizes the simplicities which would arise if there were a large number of colors. To distinguish between these disparate pictures, we must learn to calculate hadronic structure directly from QCD.

The distinction between these models is far from academic, for these models have very different implications for hadronic matter under extreme conditions. For example, in the MIT bag model in which a nucleus is viewed as being composed of large nearly overlapping bags of quarks, percolation and deconfinement are expected with only modest increases in density, whereas meson-nucleon field theory or hybrid bag models with a very small bag radius could retain hadronic degrees of freedom over a much larger density regime. Thus, timely astrophysical questions, such as supernova collapse and the composition and properties of neutron stars depend crucially on these issues, as do the major experimental initiatives to study matter under extreme conditions in the laboratory using ultra-relativistic heavy ion collisions.

These fundamental questions of hadronic structure involve exceedingly difficult many-body problems which are presently intractable by perturbative or other conventional analytic techniques. Thus, large-scale numerical calculations exploiting state of the art computing facilities in the one gigaflop domain and emerging parallel computer architectures can play a crucial role in providing a solid, quantitative foundation for the physics of strong interactions.

This work will describe initial explorations of two aspects of hadronic physics. The next section addresses the effective hadronic description of a pedagogical confining quark model. In this study, Monte Carlo techniques enable the exact solution of the ground state of quark matter and the exact scattering phase shifts of bound states, as well as the exact solution of the effective hadronic many-body problem. Thus, in contrast to the usual experience in nuclear physics, there is no uncertainty in the investigation arising from uncontrolled many-body approximations.

The last section presents the results of exploratory lattice QCD calculations to establish the foundation for quantitative study of hadron wave functions on a lattice. In order to establish and verify the computational methods for calculating Bethe–Salpeter wave functions and quark-quark correlation functions on a lattice, these quantities were calculated in two-dimensional QCD in the weak coupling domain where they could be compared directly with continuum solutions. The fact that systematic and statistical errors could be sufficiently controlled to obtain detailed agreement with continuum solutions provides strong motivation for realistic calculations in four dimensions.

II. CONFINING QUARK MODEL

The confining quark model of Ref. [1] provides a convenient system in which to study how certain aspects of nuclear structure arise from underlying quark degrees of freedom. By studying the model in one spatial dimension, it is possible to obtain exact solutions for uniform matter described both in terms of quark and hadronic degrees of freedom and thus explore the complementarity of the two descriptions.[2]

Physically, the model may be thought of as an abiabatic limit in which for any configuration of $2N$ spinless quarks, the color fields instantaneously adjust themselves to form the lowest energy configuration in which N distinct pairs of quarks are connected by flux tubes. Mathematically, the model is defined by specifying a potential energy $v(r)$ which is an increasing function of r associated with a flux tube of length r and defining the total potential energy of the quarks to be

$$V = \min_P \{v(x_{P1}-x_{P2})+v(x_{P3}-x_{P4})+\ldots+v(x_{P(N-1}-x_{PN}\} \quad (2.1)$$

where the minimum over all permutations selects the lowest energy assignment of pairs. The theory exhibits the desired separability, confinement, and exchange symmetry and is free of Van der Waals interactions. Although finding the optimal pairing in three dimensions entails solution of a non-trivial assignment problem, in one dimension with periodic boundary conditions, the optimal pairing of consecutively labeled quarks either pairs q_{2m} with q_{2m+1} or q_{2m} with q_{2m-1} for all m. Hadrons are composed of two quarks in the simplest version of the model used here and are thus Bosons. The potential $v(x)$ is taken to be quadratic, yielding a quark density in a free nucleon proportional to $x^2 e^{-\frac{x^2}{\sqrt{2}}}$

It is instructive to compare observables calculated in this model with the results obtained from an effective Hamiltonian having only hadronic degrees of freedom. As in traditional nuclear physics, a hadron-hadron potential may be defined to reproduce the exact hadron scattering phase shifts, and the many-hadron problem is solved with this potential. We find that gross properties such as the hadronic matter binding energy as a function of density and hadron-hadron correlation function are well-reproduced by this effective Hamiltonian, whereas the quark momentum distribution, which would be probed in deep inelastic electron scattering, reflect the underlying quark structure.

The phase shifts for the scattering of a two-quark hadron from a two-quark hadron may be solved stochastically below threshold by calculating the ground state of the four quark system with the appropriate boundary condition on the distance between the centers of mass of the left two quarks and the right two quarks, and the Monte Carlo results are shown in Ref. [2]. Since the phase shifts in this energy range are evidently characterized by a scattering length and an effective range, they are easy to reproduce with a simple local potential $V_N(x) = \frac{52}{\sqrt{2\pi}}\, e^{-\frac{x^2}{2}}$. Of the two effects which occur for overlapping hadrons, diminuation of the potential interaction energy and increased kinetic energy due to the Pauli principle, the later effect dominates and the interaction is purely repulsive.

The binding energy per quark of uniform nuclear matter is shown in Fig. 1. Stochastic solution of the many-quark problem yields the solid dots, and the many-hadron problem defined by Bosons interacting via the phenomenological potential $V_N(x)$ yields the triangular points.

In order to calibrate densities in the one-dimensional model, it is useful to note that for nuclear matter $\frac{\bar{\rho}_N}{\rho_{NM}} = (1.1)^3$, where $\bar{\rho}_N$ denotes the average density in a proton of uniform density and the observed rms charge radius and $\bar{\rho}_{NM}$ is nuclear matter density. Since the average density of a hadron in the one-dimensional model is $\bar{\rho}_H = 0.5$, a corresponding degree of overlap obtains at a hadronic matter density $\bar{\rho}_{HM} = \frac{0.5}{1.4} \sim 0.45$. Thus, we observe from Fig. 1 that up to densities twice that of hadronic matter, the binding energy yields no signature of quark substructure or, equivalently, of the breakdown of the description in terms of hadronic degrees of freedom.

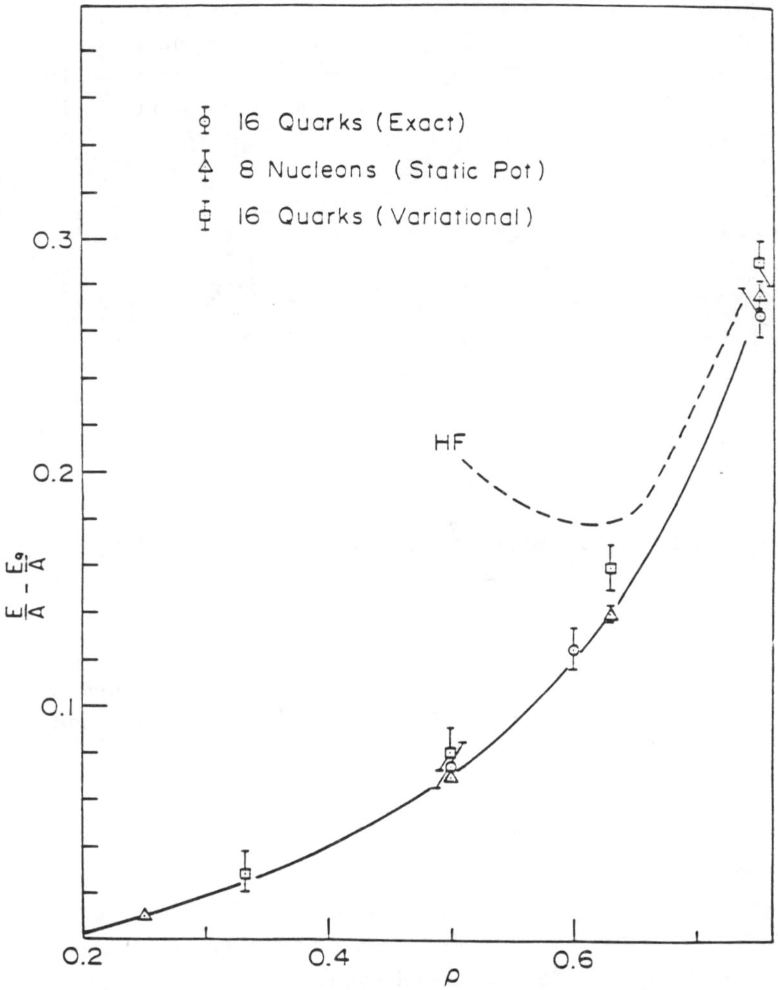

Fig. 1: Monte Carlo calculation of the binding energy per quark of uniform matter calculated with the quark Hamiltonian (circles) and the effective hadron Hamiltonian (triangles).

A second validation of the effective hadronic theory in the regime of nuclear matter density is provided by the hadron-hadron correlation function shown in Fig. 2.[3] In this model, an obvious definition of the hadron coordinates for each configuration is provided by the centers of mass of each pair of quarks connected by flux tubes, and the

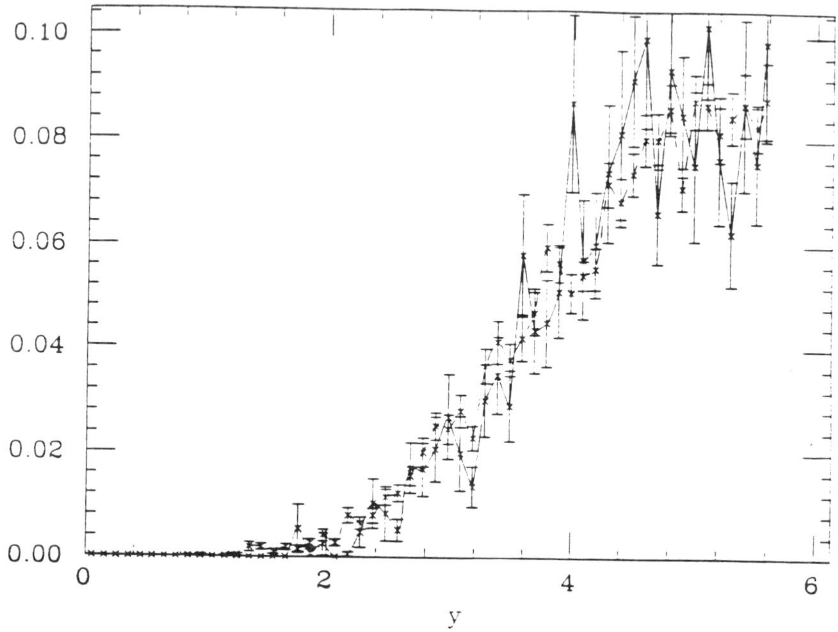

Fig. 2: Hadron-hadron correlation function calculated in the effective hadron theory (error bars connected by solid lines) and defined by the center of mass of quarks connected by flux tubes (error bars connected by dashed lines).

correlation function for these coordinates is evaluated in the quark ground state and compared with the hadron correlation function calculated in the ground state of the effective potential theory. We observe that the effective local potential very accurately reproduces the detailed spatial dependence of the correlation function. In the context of this model, the phenomenological "hard core" potential works very well indeed.

From these successes of the hadronic theory, one might naively conclude that the wave function for the quarks may be written as the antisymmetrized product of the effective hadron wave function $\Phi_{H/N}$ for the center of mass of each pair of quarks in nuclear matter times the ground state wave function for the quarks in an isolated

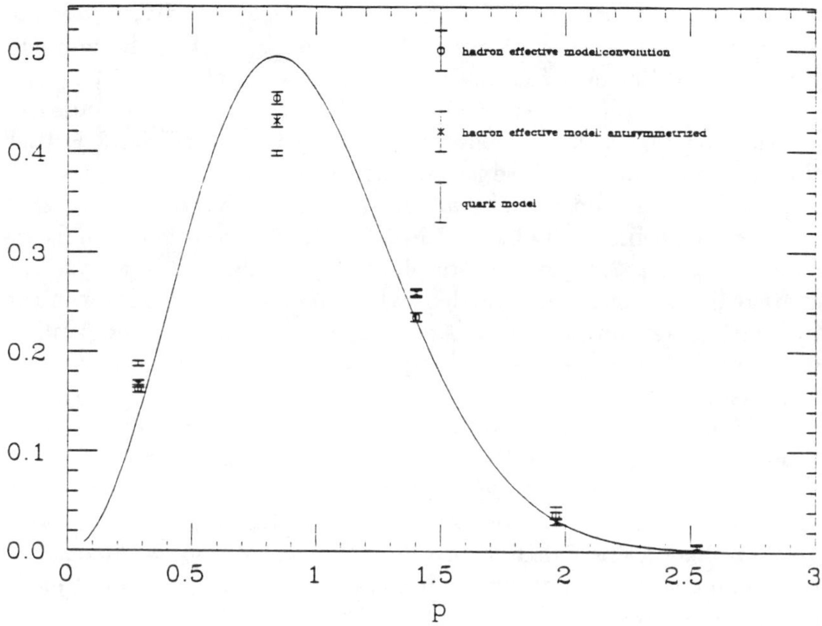

Fig. 3: The exact momentum distribution in hadronic matter (error bars), the approximation Eq. (2.2) (crosses), the convolution approximation (circles) and the momentum distribution in a free hadron (solid curve).

hadron, $\phi_{q/H}$

$$\psi(r_1 \ldots r_n) = \mathcal{A}\Phi_{H/N}\left(\frac{r_1+r_2}{2}, \frac{r_3+r_4}{2}, \frac{r_{n-1}+r_n}{2}\right) \times \phi_{q/H}(r_1 - r_2)\ldots\phi_{q/H}(r_{n-1}-r_n) \; . \tag{2.2}$$

A test of this approximation, relevant to the analysis of the EMC effect in deep inelastic lepton scattering, is to calculate the exact quark momentum distribution by stochastic evaluation of the one-body density matrix

$$n(p) = \int dy\, e^{ip\cdot y} \frac{\int dx \, \langle \phi \,| e^{-\beta H} |\, x + y \rangle \, \langle x \,| e^{-\beta H} |\, \phi \rangle}{\langle \phi \,| e^{-2\beta H} |\, \phi \rangle} \quad . \tag{2.3}$$

Monte Carlo calculations[3] for the exact momentum distribution of hadronic matter at $\rho = .35$ are compared in Fig. 3 with the momentum distribution in a free hadron, the approximation Eq. 2.2, and the cruder convolution approximation, which neglects the antisymmetrization in Eq. (2.2). The first observation from Fig. 3 is that the difference between the exact momentum distribution of quarks in hadronic matter and in free hadrons is qualitatively similar to that observed experimentally in the EMC effect. The results in the figure correspond to 2/3 nuclear matter density, roughly comparable to the average density in a heavy nucleus when the fraction of matter in the surface is taken into account. As in the EMC effect, the momentum distribution is enhanced at high and low momentum and depleted at intermediate momentum.

As expected, the convolution approximation seriously underestimates the magnitude of the effect. However, although the antisymmetrized approximation, Eq. (2.2), improves the approximation, it still is qualitatively in error. Although the present model is grossly oversimplified, this calculation raises legitimate doubts concerning the adequacy of the antisymmetrized convolution approximation in providing a quantitative description of the EMC effect.

This simple model thus embodies many features observed in low energy nuclear physics. It admits a hadronic approximation which is quantitatively accurate in the regime of nuclear matter density for observables such as the binding energy and hadron-hadron correlation function, while at the same time displaying a clear signature of quark substructure in the quark momentum distribution.

III. MESON STRUCTURE IN QCD_2

As a prelude to realistic calculations of hadronic structure in lattice QCD, a preliminary study of meson wave functions has been carried out in two space-time dimensions.[4] One should recognize at the outset a number of significant differences between QCD in two and four dimensions. The γ matrix structure in two dimensions is much simpler, giving rise for example to the essential simplifications rendering the Schwinger model solvable. Whereas there is no scale in massless four-dimensional QCD, and thus loop integrals are required to set the mass scale, in two dimensions the mass scale is already set by the coupling constant with dimension of inverse length. Similarly, confinement, which is non-trivial in four dimensions, arises trivially in two dimensions from the linear Coulomb potential and provides the foundation for the perturbative weak coupling solution. The non-Abelian aspect of the two-dimensional theory, while formally present in the lattice action, may be effectively removed in analytic treatment by the choice of axial gauge. Finally, the physics associated with chiral symmetry in four dimensions is missing in the two-dimensional counterpart both by virtue of the Mermin–Wagner theorem and the weak coupling limit we address.

WEAK COUPLING SOLUTION

Quantum chromodynamics in one space and one time dimension is defined by the Lagrangian

$$\mathcal{L} = -\frac{1}{4} F^a_{\mu\nu} F^{\mu\nu}_a + \bar{\psi}\left(i\gamma_\mu D^\mu - m\right)\psi$$

where

$$\begin{aligned} F^a_{\mu\nu} &= \partial_\mu A^a_\nu - \partial_\nu A^a_\mu + g f^{abc} A^b_\mu A^c_\nu \\ D_\mu &= \partial_\mu - ig\frac{\lambda^a}{2} A^a_\mu \\ \left[\frac{\lambda^a}{2}, \frac{\lambda^b}{2}\right] &= i f^{abc}\frac{\lambda^c}{2} \quad . \end{aligned} \tag{3.1}$$

and λ and f denote the representation matrices and structure constants for $SU(N)$.

Mesons in QCD_2 have been studied in two limits. In the large N limit, 't Hooft[5] solved the meson spectrum in the color singlet sector using light-cone variables. Subsequent work established the same result in a general axial gauge[6] and the meson wave functions have been calculated numerically.[7] Because of the increase in the number of gauge field degrees of freedom with N and the corresponding computational resources required, we have not explored this large N limit numerically.

A more convenient limit for this present investigation is the weak coupling limit, $\frac{g}{m} \to 0$. By extending Coleman's weak coupling solution of the massive Schwinger model[8] to QCD_2, Hanson *et al.*[9] have diagonalized the Hamiltonian in the one-quark, one-antiquark sector of the Fock space, and shown that this result is equivalent to the Bethe–Salpeter equation derived using standard Feynman rules with a principal value infrared regulator.

The equation of motion for the gauge field is

$$\partial_\mu F_a^{\mu\nu} = -g\left(J_a^\nu - f^{abc} F_b^{\mu\nu} A_\mu^c\right) \tag{3.2}$$

where

$$J_a^\nu = :\bar{\psi}\gamma^\nu \frac{\lambda^a}{2}\psi: \quad . \tag{3.3}$$

The theory is greatly simplified by choosing the axial gauge $A_1^a(x) = 0$, which eliminates the gauge field self-interaction and yields a static linear Coulomb-like potential. In this gauge, Eq. (3.2) becomes

$$\partial_1^2 A_a^0(x) = g J_a^0(x) \quad . \tag{3.4}$$

Integration yields the general solution

$$A_a^0(x) = \frac{g}{2}\int_{-\infty}^{\infty} dy\, |x-y|\, J_a^0(y) + xF_a(t) + B_a(t) \tag{3.5}$$

As discussed by Bars and Green[6], $B_a(t)$ can be removed by a gauge transformation and the background field $F_a(t)$ may be set to zero in the color-singlet quark-antiquark sector. The Hamiltonian may therefore be written

$$H = H_0 + H_I \tag{3.6a}$$

where

$$H_0 = \int dx{:}\bar{\psi}\left(m - i\gamma^1\partial_1\right)\psi{:} \tag{3.6b}$$

$$H_I = -\frac{g^2}{4}\int dx\, dy\, J_a^0(x)|x-y|J_a^0(y) \tag{3.6c}$$

In the weak coupling limit, it is convenient to write $\psi(x)$ in the free particle basis

$$\psi(x) = \frac{1}{\sqrt{2\pi}}\int_{-\infty}^{\infty} dp\left[A(p)u(p)e^{-ipx} + B^\dagger(p)v(p)e^{ipx}\right] \tag{3.7}$$

and to project onto the space of one quark–one anti-quark states

$$|q\bar{q}\rangle = \int_{-\infty}^{\infty} dQ\, dq\Phi(Q,q)A_\ell^\dagger(q)B_\ell^\dagger(Q-q)|0\rangle \tag{3.8}$$

As shown in Ref. [4], the eigenvalue equation in the center of mass frame ($Q = 0$) is

$$M\Phi(q) = 2\left(E(q) - \frac{G^2}{2\pi}\frac{1}{E(q)}\right)\Phi(q)$$
$$-\frac{G^2}{2\pi}P\int dp\frac{1}{(p-q)^2}\left[\frac{E(p)E(q)+pq+m^2}{2E(p)E(q)}\right]\Phi(p)\ . \tag{3.9a}$$

where

$$G^2 \equiv \frac{g^2}{2}\frac{(N_c^2-1)}{N_c} \equiv \frac{g^2 N_g}{2N_c} \tag{3.9b}$$

N_g denotes the number of generators, and M denotes the meson mass.

This equation is solved using the Multhopp technique commonly used in aerodynamics. The infinite domain $-\infty < q < \infty$ is mapped onto the interval $0 < q < \pi$ by the transformation

$$\frac{q}{m} = -\frac{\cos\theta}{\sin\theta} \tag{3.10a}$$

and $\Phi(\theta)$ is expanded as a truncated sine series

$$\Phi(\theta) = \sum_{n=1}^{N} a_n \sin n\theta \tag{3.11}$$

with the result that Eq. (3.9) becomes

$$\epsilon\sum_{n=1}^{N} a_n \sin n\theta = \sum_{n=1}^{N} B_n(\theta)a_n \tag{3.12}$$

where $B_n(0)$ is given explicitly in Ref. [4].

Finally, this equation is converted into a standard $N \times N$ eigenvalue problem by requiring that the equality hold at each of the N angles

$$\theta_k = \frac{k\pi}{N+1}\ , \qquad k = 1, 2, \ldots, N \tag{3.13}$$

From the resulting eigenvectors, it is straightforward to construct coordinate space wave functions.

SOLUTION ON A LATTICE

The lattice theory is defined by the standard Wilson action

$$S = \beta \sum_{\square} \left[1 - \frac{1}{N_c} \mathrm{Re\ Tr} \left(U_{\square} \right) \right] + \sum_{n,m} \bar{\psi}_n K(n,m) \psi_m \qquad (3.14)$$

where

$$\beta = \frac{2N_c}{g_0^2} \quad ,$$

$$K(n,m) = \delta_{nm} - \lambda \sum_{\mu} [(r - \gamma_\mu) U_\mu(n) \delta_{n+\mu,m} + (r + \gamma_\mu) U_\mu^\dagger(m) \delta_{n-\mu,m}] \quad ,$$

$$\lambda = \frac{1}{2(2r + m_0)} \quad ,$$

$U_{\square}$ denotes the path ordered product of link variables $U_\mu(n)$ on a plaquette, N_c is the number of colors, g_0 is the bare coupling constant, m_0 is the bare quark mass, and the parameter r controls the energy of spurious lattice fermion modes. Note that we will consistently distinguish parameters on the lattice from their continuum counterparts by the subscript zero, except for the parameter β which is defined only on the lattice.

To study the structure of meson bound states, we evaluate gauge-invariant operators from which the Bethe–Salpeter wavefunction and intrinsic density may be extracted. The basic idea is to use evolution in imaginary time to filter out the ground state of the QCD vacuum or a meson and sandwich appropriate operators between these states. The actual expression we evaluate is

$$O_\phi(x,t) = \sum_y \langle \bar{\psi}_u(y,t) U_1(y \to x+y,t) \gamma_5 \psi_d(x+y,t) \times \bar{\psi}_d(0,0) \gamma_5 \psi_u(0,0) \rangle \qquad (3.15)$$

where $U_1(y \to x+y,t) = \prod_{n=y}^{x+y} U_1(n,t)$, u and d are flavor labels, and $\langle \sigma \rangle = \int d\bar{\psi} d\psi dU\, e^{-S(\bar{\psi},\psi,U)} \sigma / \int d\bar{\psi} d\psi dU\, e^{-S(\bar{\psi},\psi,U)}$ denotes the integral over all quark and gauge fields weighted with the action for a lattice with total time extent T. The physical content of $\mathcal{O}_\phi(x,t)$ is seen by inserting complete sets of states $\sum_\alpha |\alpha\rangle\langle\alpha|$.

$$\mathcal{O}_\phi(x,t) = \sum_{\alpha,\beta}\sum_y \langle\beta|\bar{\psi}_u(y)U_1(y\to x+y)\gamma_5\psi_d(x+y)|\alpha\rangle \\ \times \langle\alpha|\bar{\psi}_d(0)\gamma_5\psi_u(0)|\beta\rangle\, e^{-M_\alpha t}e^{-M_\beta(T-t)}\left(\sum_\gamma e^{-M_\gamma T}\right)^{-1} \tag{3.16}$$

In the limit in which the time extent of the lattice, T, is sufficiently large, the dominant state in the β sum is the vacuum $|\beta\rangle = |0\rangle$, and $\mathcal{O}_\phi$ may be rewritten

$$\mathcal{O}_\phi(x,t) = \sum_\alpha \phi_\alpha(x)C_\alpha e^{-M_\alpha t} \tag{3.17}$$

where

$$\phi_\alpha(x) = \sum_y \langle 0|\bar{\psi}_u(x)U_1(y\to x+y)\gamma_5\psi_d(x+y)|\alpha\rangle \quad . \tag{3.18}$$

is the Bethe–Salpeter wavefunction for the meson $|\alpha\rangle$, C_α is an overall normalization constant proportional to the wavefunction at the origin, and the mass M_α is measured relative to the energy of the vacuum. The integration over y projects onto states $|\alpha\rangle$ with zero momentum and the distinct color labels are introduced to reduce the number of fermion contractions. Note that this definition of the Bethe–Salpeter wavefunction includes the product of link variables required to make $\mathcal{O}_\phi(x,t)$ gauge invariant. For a finite lattice, the periodic boundary condition prevents setting one spatial link variable to unity, and thus introduces a finite boundary effect. The technical details associated with the T and t dependence, boundary effects, and lattice artifacts are discussed in the Appendices of Ref. [4].

To evaluate the intrinsic density distribution for a meson state $|\alpha\rangle$, we need to calculate the density-density correlation function

$$\rho_\alpha(x) = \sum_y \langle\alpha|j_0(y)j_0(x+y)|\alpha\rangle \tag{3.19}$$

where again $|\alpha\rangle$ is prepared by filtering by evolution in imaginary time.

The observable analogous to $\mathcal{O}_\phi(x,t)$ which projects out two meson states with two density operators between them is

$$\begin{aligned}
\mathcal{O}_\rho(x,t,t') &= \sum_y \big\langle \bar{\psi}_u(0,t)\gamma_5\psi_d(0,t)j_u^0(y,t')j_d^0(x+y,t') \\
&\qquad \times \bar{\psi}_d(0,0)\gamma_5\psi_u(0,0)\big\rangle \\
&= \sum_{\alpha,\beta,\gamma} \big\langle\gamma|\bar{\psi}_u(0)\gamma_5\psi_d(0)|\alpha\big\rangle\,\rho_{\alpha\beta}(x) \\
&\qquad \times \big\langle\beta|\bar{\psi}_d(0)\gamma_5\psi_u(0)|\gamma\big\rangle\,e^{-(t-t')M_\alpha}e^{-t'M_\beta} \\
&\qquad \times e^{-(T-t)M_\gamma}\times\left(\sum_\delta e^{-M_\delta T}\right)^{-1}
\end{aligned} \tag{3.20}$$

where we have defined $\rho_{\alpha\beta}(x)=\sum_y\big\langle\alpha|j_u^0(y)j_d^0(x+y)|\beta\big\rangle$. The diagonal elements of $\rho_{\alpha\beta}(x)$ are the density-density correlation functions.

To understand the relationship between $\rho_{\alpha\beta}(x)$ and $\phi_\alpha(x)$, consider the weak coupling limit where the meson wavefunction has the form (3.8). Evaluation of the matrix element of the operator $\sum_y j_u^0(y)j_d^0(x+y)$ in this meson state yields

$$\begin{aligned}
\sum_y \big\langle u\bar{d}|j_u^0(y)j_d^0(x+y)|u\bar{d}\big\rangle &= \frac{1}{2\pi}\int dk\,dp\,dk'\,dp' u^\dagger(p)u(k)v^\dagger(p')v(k') \\
&\times \delta(p-k-p'+k')e^{-i(p'-k')x}\Phi(k'+p,p)\Phi(p'+k,k)
\end{aligned} \tag{3.21}$$

In the non-relativistic limit, in which $u(p)\to\begin{pmatrix}1\\0\end{pmatrix}$ and $v(p)\to\begin{pmatrix}0\\1\end{pmatrix}$, Eq. (3.21) reduces to a normalization constant times $\phi^2(x)$ so that the intrinsic density is the square of the relative wave function as expected. In general, however, $\rho(x)$ is not proportional to $\phi^2(x)$ because of the factor $u^\dagger(p)u(k)v^\dagger(p')v(k')$. This factor corresponds to the product of form factors at each density operator vertex, where there is a momentum transfer $q=p-k=p'-k'$, and the spinor overlaps yield $u^\dagger(k+q)u(k)$ and $v^\dagger(k'+q)v(k')$ which are 1 at $q=0$ and decrease as q increases. If we write $\rho(x)$ schematically as a Fourier transform $\rho(x)=\int dq\,e^{iqx}\Phi^2(q)F(q)$ we expect $\rho(x)$ to behave as the convolution of $\phi^2(x)$ with a finite density distribution and thus decay more slowly than $\phi^2(x)$, and this expectation is borne out by the evaluation of (3.21) presented below.

The regime of parameters explored in this work was determined by the following considerations. Recall there are three dimensionful parameters in the lattice theory, the lattice spacing with dimension of length $[a] = L$, the bare mass $[m_0] = L^{-1}$, and the bare coupling constant $[g_0] = L^{-1}$, or equivalently the string tension in the weak coupling limit $\sigma_0 = \frac{G_0^2}{2} \equiv \frac{N_g}{2\beta} \equiv \frac{N_g g_0^2}{4N_c}$ with $[\sigma_0] = L^{-2}$. We wish to simultaneously satisfy several conditions summarized in Table I. The continuum limit in the pure gauge sector requires $a^2\sigma_0 << 1$. Operationally we used the criterion that the string tension be dominated by the lowest non-trivial order in the weak coupling expansion,[10]

$$\sigma_0 = -\ln W(\Box) = \frac{N_g}{2\beta} + O(\beta^{-2}) \tag{3.22}$$

TABLE 1: Constraints on parameters. The physical constraints on dimensionless combinations of parameters and the values of these combinations in $SU(2)$ with $m_0 = 1/2$ and $a = 1$ for two values of β are tabulated.

CONSTRAINT		$\beta = 12$	$\beta = 48$
Gauge Field Continuum Limit	$a^2\sigma_0 < 1$	$\frac{1}{8}$	$\frac{1}{32}$
Fermion Continuum Limit	$am_0 < 1$	$\frac{1}{2}$	$\frac{1}{2}$
Lattice Sites within Meson	$1 < \frac{1}{a(m_0\sigma_0)^{1/3}} < 7$	2.5	4
Weak Coupling and Non-Relativistic limit	$\frac{\sigma_0}{m_0^2} < 1$	$\frac{1}{2}$	$\frac{1}{8}$

which occurs in our numerical results for $\sigma_0 = \frac{N_g}{2\beta} < \frac{1}{8a^2}$. The continuum limit for Fermions requires that the characteristic Fermion momentum must be much less than $\frac{1}{a}$. In the relativistic regime, the Compton wavelength gives the stringent constraint $m_0 a << 1$ whereas the characteristic momentum of the non-relativistic bound state, $p \sim \frac{1}{L_{\rm NR}} = (m\sigma)^{1/3}$ requires only $a\,(m_0\sigma_0)^{1/3} << 1$ which is satisfied for $am_0 \sim 1$ as long as $(a^2\sigma_0) << 1$. To describe the spatial distribution of a bound state accurately, we must have enough lattice sites within the spatial extent of the meson, $\frac{L_{\rm NR}}{a} > 1$ while still having $\frac{L_{NR}}{a}$ much less than the number of spatial mesh sites. In our calculations we could conveniently include 14 lattice sites, so these two requirements lead to the condition $1 < \frac{1}{a(m_0\sigma_0)^{1/3}} < 7$. Finally, the weak coupling and non-relativistic limits both require $\frac{\sigma}{m^2} << 1$. From the numerical experience of Ref. [9], we know that the weak coupling expansion is good for $R > 2$ or $\frac{\sigma}{m^2} < \frac{\pi}{4}$. Combined with the requirement $\sigma_0 < \frac{1}{8a^2}$, this led us to choose $m_0 = \frac{1}{2a}$ and for convenience, we use units such that $a = 1$. Table I shows how this choice of parameters simultaneously satisfies all these conditions for the range of $SU(2)$ couplings we studied with β ranging from 11 to 48.

The Monte Carlo calculations in this work are performed in the quenched approximation, in which the Fermion determinant is ignored in sampling the action. The errors arising from this approximation become negligible in the weak coupling regime where the contributions of quark loops are suppressed. We used the heat bath method of Cabibbo and Marinari[11] for $SU(N)$ gauge field updating and the conjugate gradient method for calculating Fermion propagators.

A typical Monte Carlo run is illustrated by the case of $SU(2)$ with $\beta = 12.0$ and $m_0 = 0.5$, for which we used a lattice size of 18×14. After 200 thermalization sweeps, 100 independent gauge field configurations, each separated by 20 MC sweeps, were used to calculate the ensemble average. The CPU time for such a run is less than 10 hours on a VAX 780. The Wilson parameter r is usually set to be 1.0 unless otherwise explicitly stated. The lattice size is chosen such that the Bethe–Salpeter amplitude decays by one order of magnitude within the finite lattice.

RESULTS

The principal difficulty in lattice calculations is obtaining a controlled and convincing approximation to the continuum limit. A necessary, although not sufficient, condition is the correct scaling behavior. In the present case, for any value of the lattice spacing, a, we may determine the bare parameters of the theory m_0 and $\beta = \frac{2N_c}{g_0^2} = \frac{N_g}{G_0^2}$ by reproducing two physical observables which we choose to be the meson mass M_π and the physical string tension σ_0 specifying the force law between two static charges measured by a Wilson loop. As a becomes sufficiently small, we expect other

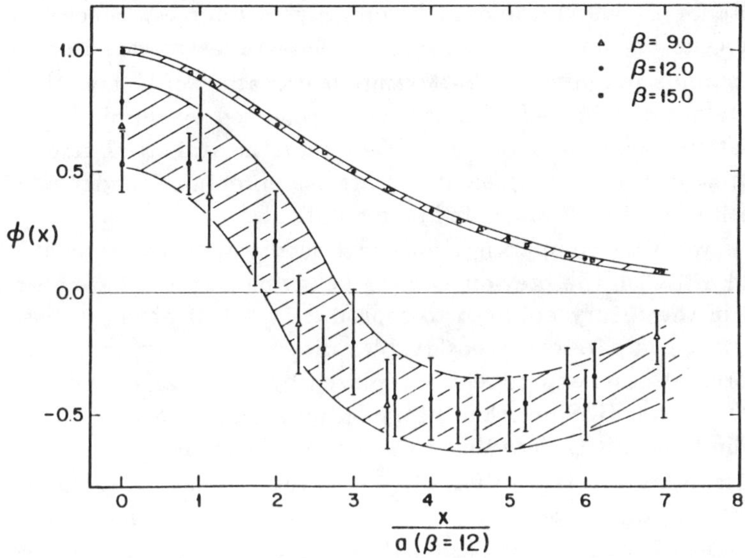

Fig. 4: Scaling behavior of the Bethe–Salpeter wavefunction for the meson states. The ground state is normalized to one at the origin, the even excited state is normalized at the outer maximum, the shaded regions and error bars indicate the statistical Monte Carlo errors, and x is measured in units of the lattice spacing for $\beta = 12$.

physical observables, such as the Bethe–Salpeter wavefunction, to approach a unique limit independent of a.

Figure 4 shows that lattice calculations of the Bethe–Salpeter wavefunction are consistent with such a limit. In these calculations, we picked a value of β as the independent variable and calculated a single Wilson loop $W(\square)$ to specify the lattice spacing a in terms of the fixed string tension σ_0 according to

$$a^2(\beta)\sigma_0 = -\ln W(\square) \tag{3.23}$$

Note that in two dimensions, an arbitrary Wilson loop on an infinite lattice or finite lattice with free boundary conditions factorizes into the product of single Wilson loops for each of the enclosed plaquettes, so that Eq. (3.23) assumes that the lattice is large enough that boundary effects are negligible. The bare mass, m_0, is adjusted to obtain the fixed meson mass M_π. To within statistical errors, the results for $\beta = 9$, 12, and 15 lie on universal curves. Since in the weak coupling limit, $\sigma_0 \sim \frac{N_g}{2\beta a^2}$, the 50% range of variation of β corresponds roughly to a 25% range of variation of a, and the actual values of a in each calculation are reflected by the positions of the lattice points in the figure. Note that the statistical errors associated with extracting excited state wavefunctions render precise quantitative studies of excited states difficult.

If we were comparing a physical theory with experiment, we would establish the previous scaling behavior with the bare parameters of the theory adjusted to reproduce a set of physical masses, and then compare other observables directly with experiment. In the present case, experiment is replaced by the weak coupling theory of Section II. Since this theory only becomes exact in the weak coupling limit, it is clear that we must make a series of comparisons at successively weaker values of $\frac{G^2}{2m^2}$ to test the lattice calculation. In order to minimize the computational effort, instead of fine tuning a set of Monte Carlo calculations to compare with each weak coupling solution, for each Monte Carlo calculation with fixed bare parameters, we prefer to compare with an appropriate continuum solution corresponding to that set of bare parameters. To do so, we must choose two conditions to relate the two parameters of the lattice theory, which we take as G_0^2 and m_0 with $a = 1$, and the two parameters of the weak coupling continuum theory, G^2 and m.

One natural constraint on the parameters is the mass of the pseudoscalar meson, so we require that the weak coupling meson mass, $M_\pi^{\mathrm{WC}}(G^2, m)$, equal the meson mass in the lattice theory $M_\pi^{\mathrm{L}}(G_0^2, m_0)$. A second possible constraint is the mass of the scalar meson, M_s, which is defined analogous to $\mathcal{O}_\phi(x,t)$ in Eq. (3.15) with $\bar\psi\gamma_5\psi$ replaced by $\bar\psi\psi$. As in the case of Fig. 4, we can solve the weak coupling theory for the dependence of $M_s^{\mathrm{WC}}(G^2, m)$ on G^2 and m. Figure 5 shows the constraints obtained for the case of a Monte Carlo calculation of M_π^{L} and M_s^{L} with $G_0^{-2} = 4$ and $m_0 = 0.5$. Ideally, we would like to show two curves, one specifying the set of values of the weak coupling parameters m and G^{-2} such that $M_\pi^{\mathrm{WC}}(G^2, m) = M_\pi^{\mathrm{L}}$ and a second specifying the values such that $M_s^{\mathrm{WC}}(G^2, m) = M_s^{\mathrm{L}}$. The intersection of these curves would then uniquely determine the values of G^2 and m for which to compare other observables such as the detailed spatial dependence of Bethe–Salpeter wavefunction and density-density correlation function. Because of the large Monte

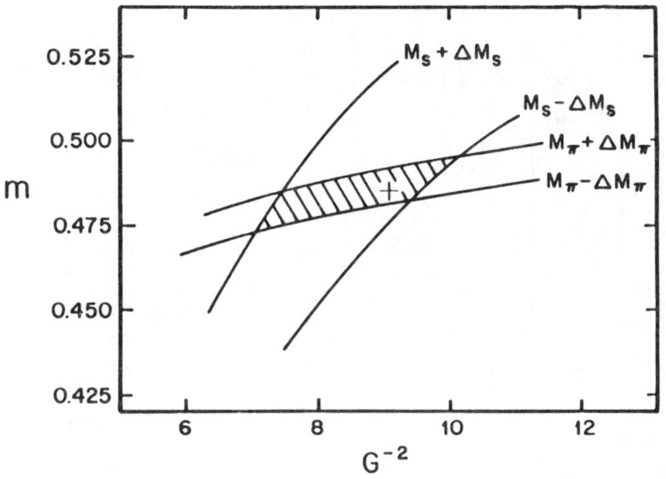

Fig. 5: Constraints on m and G^{-2} specified by fixing the pseudoscalar meson mass M_π and scalar meson mass M_s. The curves indicate the constraints obtained at the upper and lower limits of the Monte Carlo error bars, and the shaded region denotes the range of parameters consistent with the Monte Carlo constraints.

Carlo errors, we have actually shown two curves for each mass constraint, corresponding to the upper and lower limits of the error bars, so that the shaded region defines the acceptable range of m and G^2. The shift in m from $m_0 = 0.5$ is quite small and well-determined. Although the result $G^{-2} = 8.3 \pm 1.3$ obtained from the figure shows that G^{-2} is shifted appreciably from the bare value $G_0^{-2} = 4$, the error in this determination of G^{-2} is inconveniently large for quantitative comparisons. In Ref. [4], a more convenient constraint is described, which for the present purposes, should be viewed as equivalent to the result in Fig. 5.

The comparison between the weak coupling and lattice solutions for the ground state pseudoscalar meson Bethe–Salpeter wavefunctions is shown in Fig. 6. The lattice Monte Carlo results are in excellent agreement with the solid lines, corresponding to weak coupling solutions with G^2 and m determined from the Monte Carlo meson mass M_π and M_s. For comparison, the weak coupling solutions with the naive correspondence $G^2 = G_0^2$ are also shown by the dashed line, and even this limit is approached in the ultra-weak coupling regime.

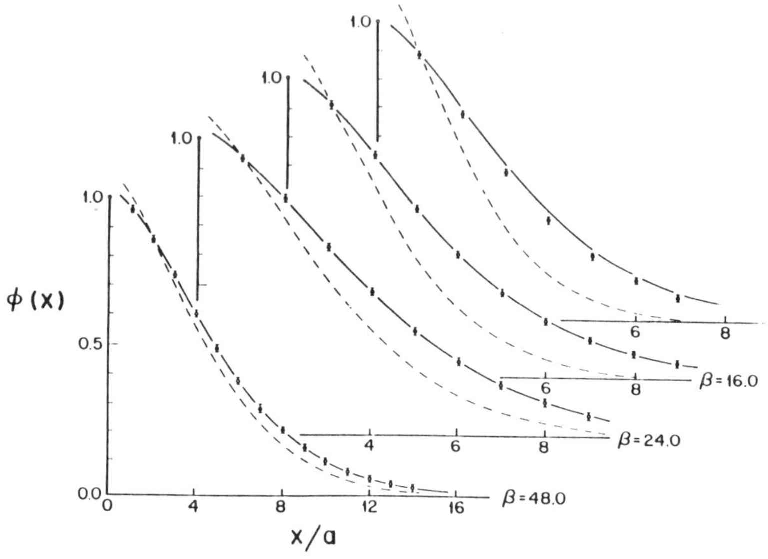

Fig. 6: Bethe–Salpeter wavefunctions for the pseudoscalar meson ground state. The error bars denote $SU(2)$ Monte Carlo calculations with $m_0 = 0.5$ and the indicated values of $\beta = \frac{N_g}{G_0^2}$. The solid lines denote the weak coupling solutions with G^{-2} and m specified by the meson masses M_π and M_s. The dashed lines show the weak coupling solutions resulting with $G^2 = G_0^2$ and only the mass m fit to reproduce M_π.

As emphasized in above, the current-current correlation function corresponds to a physical observable and provides information complementary to the Bethe–Salpeter amplitude. Since the Bethe–Salpeter wavefunction is restricted to the quark-antiquark sector of the Fock space while the density operator measures contributions from all sectors, differences between these two quantities can provide important insight into hadron structure. In the present model, where the differences are clearly understood in the weak coupling regime, our objective is to show that both quantities and their differences can be studied quantitatively in the lattice theory.

For the ground state of the pseudoscalar meson $|\pi\rangle$, the density-

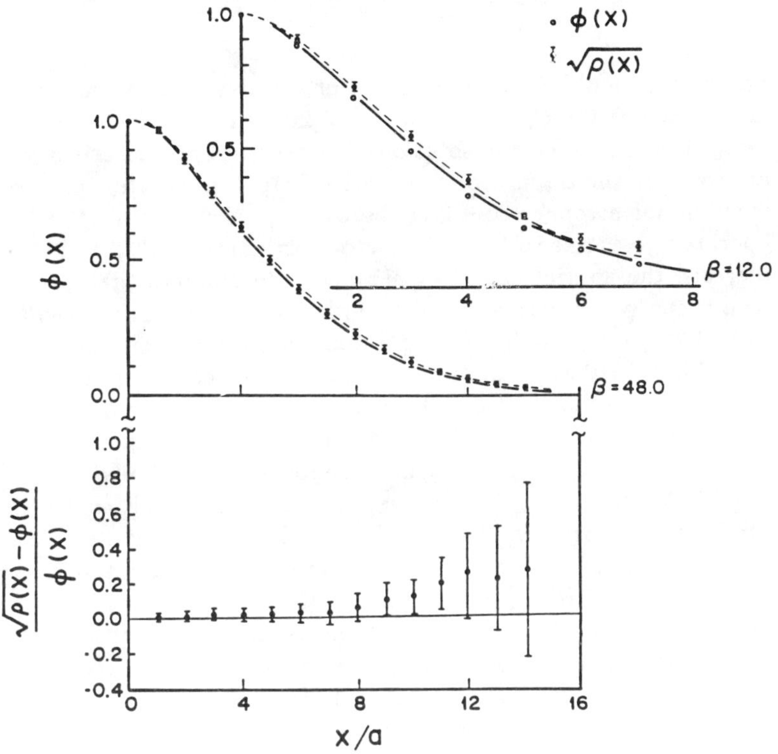

Fig. 7: Comparison of the density-density correlation function and Bethe–Salpeter wavefunction. Monte Carlo results for the square root of the density-density correlation function $\sqrt{\rho(x)}$ are denoted by the solid points with error bars representing statistical errors and the weak coupling solution is given by the dashed lines. The Bethe–Salpeter results for $\phi(x)$ from Fig. 6 are also shown for comparison, with the weak coupling solutions given by the solid line and the Monte Carlo results for $\beta = 12$ given by the open circles, the size of which are comparable to the statistical errors. Since the Monte Carlo results for $\sqrt{\rho(x)}$ and $\phi(x)$ at $\beta = 48$ cannot be distinguished in the scale of the figure, the fractional deviation between them and the corresponding statistical errors are shown on the lower curve. Note that all results were normalized to one at the origin.

density correlation function

$$\rho(x) = \sum_{y} \langle \pi \left| j_u^0(y) j_d^0(x+y) \right| \pi \rangle \tag{3.24}$$

has been calculated as a function of relative separation using the lattice expression $\mathcal{O}_\rho(x,t,t')$ in Eq. (3.20). In the extreme weak coupling limit in which the solution also becomes non-relativistic, $\rho(x)$ approaches the square of the Bethe–Salpeter wavefunction $\phi^2(x)$, whereas for stronger coupling the overlap of spinors at the density operator gives rise to the form factor corrections in Eq. (3.21) which increase the spatial extent of $\rho(x)$. Figure 7 shows that the lattice results for $\sqrt{\rho(x)}$ agree precisely with the weak coupling result from Eq. (3.21), denoted by dashed lines. for comparison, the previous Bethe–Salpeter wavefunction, $\phi(x)$, and the weak coupling solution, given by the solid line, are also shown. The approach of $\sqrt{\rho(x)}$ and $\phi(x)$ at weak coupling and the known small discrepancy at stronger coupling are clearly and accurately reproduced by the lattice results.

This study of QCD_2 in the weak coupling limit in which the continuum theory may be solved numerically demonstrates the feasibility of using lattice calculations as a quantitative tool to study hadronic structure. The principal results are the accurate determination of the spatial dependence of the Bethe–Salpeter wavefunction and density-density correlation function in precise agreement with the weak coupling solutions shown in Figs. 6 and 7.

Clearly, many additional issues must be addressed before one can obtain comparable accuracy for physical hadrons in three spatial dimensions, such as the effect of dynamical Fermions, adequate treatment of chiral symmetry, and potential problems of cancellations in calculating baryons. Nevertheless, these initial results are encouraging and suggest that the study of QCD on a lattice may indeed provide a tool to gain insight into the structure of hadrons.

IV. CONCLUSION

This work has described initial explorations in which large scale numerical calculations are being used to study hadron structure. Clearly, there are many pitfalls in progressing from these model calculations in one spatial dimension to realistic calculations in three spatial dimensions. At this stage, we can only hope that the study of judicious combinations of quark and gluonic observables will provide fundamental insight into hadronic structure, or, in the terms of the conference title, that our future offers more gigaflops than flops.

REFERENCES

1. F. Lenz, J. T. Longergan, E. J. Moniz, R. Rosenfelder, M. Stingl and K. Yazaki, *Ann. Phys.* **170**, 65 (1986).
2. C. Horowitz, E. Moniz, and J. W. Negele, *Phys. Rev. D* **31**, 1689 (1985).
3. M. Lissia and J. W. Negele, to be published.
4. S. Huang, J. W. Negele and J. Polonyi, *Nucl. Phys. B*, in press.
5. G.'t Hooft, *Nucl. Phys.* **B75** (1975) 461.
6. I. Bars and M. B. Green, *Phys. Rev.* **D17** (1978) 537.
7. M. Li, L. Wilets and M. C. Birse, *J. Phys. G: Nucl. Phys.* **13** (1987) 915.
8. S. Coleman, *Ann. Phys.* (NY) **101** (1976) 239.
9. A. J. Hanson, R. D. Peccei and M. K. Prasad, *Nucl. Phys.* **B121** (1977) 477.
10. M. Creutz, *Phys. Rev. Lett.* **43** (1979) 553
11. N. Cabibbo and E. Marinari, *Phys. Lett.* **199B** (1982) 387.

The Relativistic Mean-Field Model in Nuclear Physics

P.-G. Reinhard, *
V. Blum, J. Fink, M. Rufa, J. Maruhn, W.Greiner, †
C. Bottcher, M.R. Strayer, S. Umar, J. Wu, ‡
and
R.Y. Cusson§

Abstract

Results from the Relativistic Mean-Field model of the nucleus for deformed nuclei and heavy-ion collisions are presented. Technical details of gridding and solving the Dirac equation are discussed.

1 Introduction

Mean-Field models employing effective interactions have been in the last two decades the key tool for microscopic models of nuclear structure and nuclear collective motion [1]. They allow to describe a wide range of phenomena in nuclear and heavy-ion physics, for example: *ground state* properties, surface and volume *vibrations*, stable *deformations* with their according *rotational* spectra, *isomeric states* and their related transitions, *fission* as a collective tunneling phenomenon, *fusion* above and below the Coulomb barrier, and *heavy-ion collisions*. In fig.1 we try to symbolize and enumerate these phenomena. They have been much studied in the past, and they still provide ample experimental information which can be exploited to test and to improve theoretical models of nuclear dynamics. In fact, the material is growing with the heavy-ion accelerators going up in energy and precision.

Most investigations work with the nonrelativistic Skyrme-Hartree-Fock model which is based on a density- and momentum-dependent zero-range

*Inst.f.Theor.Physik, Universität Erlangen, W.Germany
†Inst.f.Theor.Physik, Universität Frankfurt, W.Germany
‡Oak Ridge National Laboratory, Oak Ridge/TN, USA
§SAIC, San Francisco /CA, USA

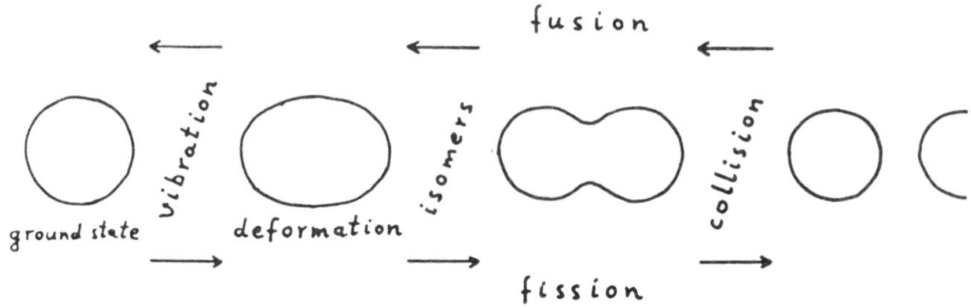

Figure 1:
Sketch of nuclear shapes in the various stages of a large amplitude collective motion. The lines illustrate the shape of the surface of the nuclear mean field. The arrows indicate the evolution of the collective motion.

interaction [2]. But in recent years, a relativistic mean-field model of the nucleus has attracted more and more interest [3,4] (although it exists already as long as its nonrelativistic counterpart). It offers the advantages of conceptual simplicity, of automatically embracing spin-properties, and of providing the necessary retardation effects for energetic processes. The studies of large amplitude collective motion with the relativistic mean-field model are still in a developing stage. Some of the first results will be presented in section 2. They are very promising and call for further investigations. These, however, are numerically demanding tasks. We will discuss two of the related numerical problems in that contribution, first, in section 3 an efficient coordinate space representation, and second, in section 4 fast iteration schemes to solve the Dirac equation.

2 The relativistic mean-field model

2.1 The coupled field equations

The relativistic mean-field model describes the nucleus as a system of nucleons which obey the Dirac equation and which are coupled via meson fields. The meson fields are treated as classical c-number fields; that is the *mean-field* approximation. And the contributions of the Dirac sea of antinucleons are neglected; that is the *no-sea* approximation. This yields the coupled equations for the nucleon single particle wave functions

$$i\frac{\partial}{\partial t}\gamma_0\varphi_\alpha = (-i\vec{\gamma}\cdot\vec{\nabla} + m_B + g_\sigma\Phi + g_\omega V^\mu\gamma_\mu + \frac{1}{2}g_\rho\vec{R}^\mu\gamma_\mu\vec{\tau} + eA^\mu\gamma_\mu\frac{1+\tau_0}{2})\varphi_\alpha \tag{1a}$$

and the meson fields

$$\Box\,\Phi + m_\sigma^2\Phi + b_2\Phi^2 + b_3\Phi^3 = -g_\sigma\textstyle\sum_{\alpha=1}^{A}\varphi_\alpha^+\varphi_\alpha \tag{1b}$$

$$(\Box + m_\omega^2)V_\mu = g_\omega\textstyle\sum_{\alpha=1}^{A}\varphi_\alpha^+\gamma_\mu\varphi_\alpha \tag{1c}$$

$$(\Box + m_\rho^2)\vec{R}_\mu = \tfrac{1}{2}g_\rho\textstyle\sum_{\alpha=1}^{A}\varphi_\alpha^+\vec{\tau}\gamma_\mu\varphi_\alpha \tag{1d}$$

$$\Box A_\mu = e\textstyle\sum_{\alpha=1}^{A}\varphi_\alpha^+\frac{1+\tau_0}{2}\gamma_\mu\varphi_\alpha \tag{1e}$$

where Φ is a scalar field (σ-meson), V_μ a vector field (ω-meson), $\vec{R}_\mu$ an isovector-vector field (ρ-meson), and A_μ the photon field. The sums for the densities may be complemented by occupation weights w_α in order to allow for the treatment of open-shell systems. Note that the *mean-field* approximation provides an independent single particle motion for the nucleons, and this, in turn, brings the densities into the simple form above. The π-meson which is usually considered to be the predominant meson cannot occur at the mean-field level because the pseudoscalar density vanishes for nuclear states of good parity.

The Lagrangian for the equations-of-motion (1) is considered to be an effective Lagrangian to be used in connection with the *mean-field* and *no-sea* approximations. The meson parameters of the relativistic mean-field model are thought to absorb all many-body and quantum field effects neglected in the approximation. For example, one expects that the *sigma*-meson is mainly to simulate two- and three-π exchange [12]. Up to now one is missing a theoretical derivation of the effective Lagrangian from some more microscopic picture of the nucleons. Thus the meson parameters g_σ, g_ω, g_ρ, m_σ, m_ω, m_ρ b_2 and b_3 are considered to be the free parameters of the model. They can be adjusted to provide an excellent reproduction of ground state properties of spherical nuclei [5,6]. The relativistic mean-field model with these optimized parameters is competitive to the well established nonrelativistic Skyrme-Hartree-Fock model. The bulk properties of the nuclear ground state can be adjusted very well. In the average over a selection of spherical nuclei from ^{16}O up to ^{208}Pb one obtains a reproduction of the binding energies E_B within 0.26%, of the diffraction radii R within 0.7%, and of the surface thicknesses σ within 3.5% [5,6]. It is to be noted that the spin-orbit force comes out just right without any special adjustment. That is one of the strengths of a relativistic treatment. However, we find that the effective mass, $m^* = m + g_\sigma \Phi$, comes out rather low. That is a topic of concern for future investigations.

2.2 Deformed calculations

A wide range of data is accessed if one extends the calculations to deformed systems. One can look for stable ground state deformations and for the structure of large amplitude collective motion. Deformations properties depend on the strength of the spin-orbit splitting as well as on the effective mass of the nucleons in the mean-field which both determine the spectral density of the single particle states. Thus they allow an indirect test of the single particle spectra of the model. That is particular welcome because those spectra are not directly accesible in experiment.

We have performed deformed calculations for light and for heavy nuclei on a coordinate space grid using finite differences for the derivatives [13,14]. The four components of the Dirac wave functions are placed on shifted grids in order to avoid the so called Fermion doubling problem. A schematic pairing force is included to allow for the treatment of nonclosed shell nuclei. A quadrupole constraint is added to access each desired deformation of the nucleus. The total energy as function of the (quadrupole) deformation gives the so called potential energy surface from which one can read-off many features of nuclear deformation and collective motion.

In fig.2 we show the potential energy surface for various even isotopes of Gd. One clearly sees the phase transition from spherical ^{146}Gd to the deformed heavier isotopes. Such a phase transition is expected in that region, and it is remarkable that the relativistic mean-field model reproduces this effect. However, there is a discrepancy in quantitative detail: the phase transition occurs earlier in the model than in experiment. The precise point of the phase transition is a very sensitive test on a model and provides selective material to discrimiate possible parametrisations. We find indeed that the isotope where the phase transition occurs and the size of the ground state deformation vary if we use slightly varied parametrisations. Much needs to be done to exploit this sensitivity.

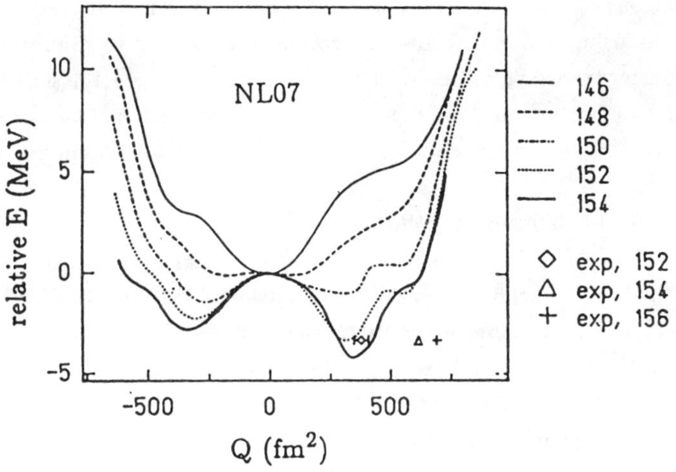

Figure 2:
Potential energy surfaces for quadrupole deformation for the even isotopes of Gd computed with a parametrisation which gives an effective nucleon mass of $m^*/m = 0.7$*, see [6]. The curves are rescaled to coincide at zero deformation. The assignment of the isotopes to the curves is given at the right hand side of the figure. The symbols indicate the available experimental ground state deformations deduced from B(E2) values.*

2.3 Time-Dependent Mean-Field calculations

Time-dependent Hartree-Fock calculations with (nonrelativistic) Skyrme forces have been widely used to describe microscopically all sorts of heavy-ion collisions [1]. Time-Dependent applications of the relativistic mean-field model are particularly interesting for energetic heavy-ion collisions where relativistic kinematics is certainly needed. Here, will report shortly on the calculations of refs. [15,16]. These have been done in full three spatial dimensions using a momentum- and coordinate-space grid connected by the fast Fourier transformation. A fifth order predictor-corrector scheme is used for the time evolution. The meson equations are split into two equations of first order in the time derivative to allow the same time stepping as for the nucleons.

Results for a head-on collision of $^{16}O +^{16} O$ at $600MeV$ per particle are shown in fig.3. In contrast to nonrelativistic calculations, the relativistic mean-field model produces complete spallation of the nuclei. This behaviour is observed also at other energies studied ($300MeV$ and $1200MeV$) and at all impact parameters. The reason for this behaviour is the fact that the mesonic fields are degrees-of-freedom on their own in a relativistic model. Due to different meson masses the attractive scalar field and the repulsive vector field get out of phase during the collision. This produces huge oscillations in the meson fields and this, in turn, causes the nuclear binding to dissolve. Those calculations show a clear relevance of a relativistic treatment in medium energy nuclear dynamics, and they call again for more investigations of that kind.

3 Comment on Basis Splines

The deformed shapes require a representation of the wave functions in two or, even better, three spatial dimension. Thus one aims at a most efficient representation in order to keep the expenses low. Most customary is a representation on a coordinate-space grid with finite difference formulas for the derivatives (which are needed e.g. in the kinetic energy). These schemes allow to work with sparse matrices which is an advantage for old-fashioned scalar computers. However, they require a rather large number of grid points because the finite difference formulas are not very precise, in particular in relativistic models where a large variation of the effective mass in the kinetic

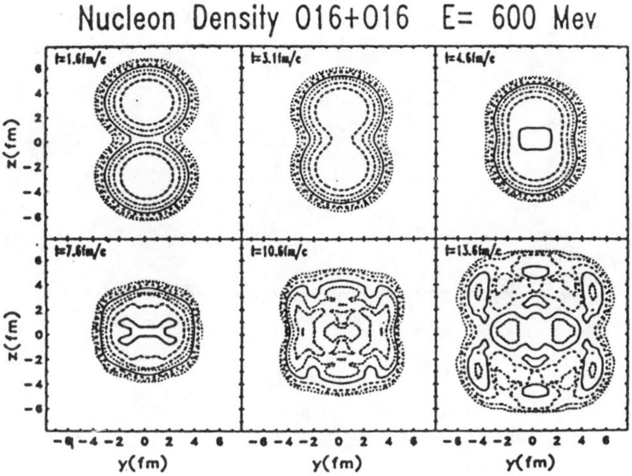

Figure 3:
Equidensity plots of the nucleon density in the y-z-plane at various stages of a head-on collsion of $^{16}O + ^{16}O$ *at* $600MeV$ *per nucleon in the laboratory frame, taken from [16].*

energy is encountered.

As an alternative, a representation in terms of Basis Spline's has been developed recently [7] which is suited especially for vector- or matrix-processors, which allows to reduce substantially the size of the grid, and which still keeps the advantages of working in a coordinate space representation. We will give a short account of it here. A detailed explanation is found in ref. [7].

A wave function $\psi(x)$ is represented in terms of basis functions

$$\psi(x) = u_\alpha^{(N)}(x)\psi^\alpha \tag{2}$$

where the ψ^α are the expansion coefficients and summation over double indices is implicitly assumed. The $u_\alpha^{(N)}(x)$ are the so called Basis Splines where the upper index indicates the order. In fig.4 we give a schematic view of a third order Basis Spline function. The Basis Splines have the following properties:

1. They are N-th. order piecewise polynomials.

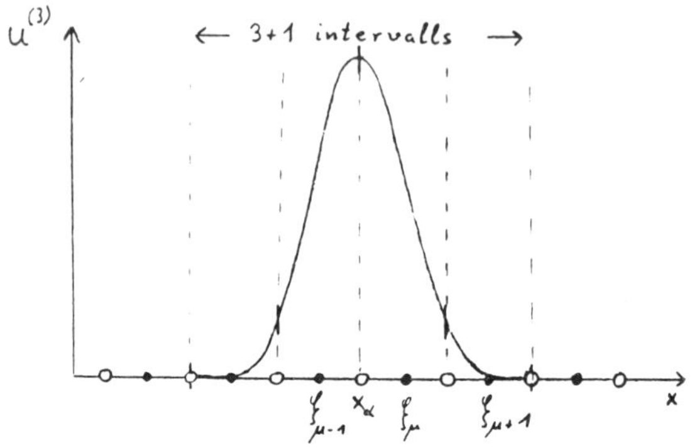

Figure 4:
Schematic plot of a third order Basis Spline $u_\alpha^{(3)}(x)$. The ∘ indicate the "knots" on the mesh where the Basis Splines are connected, and the • indicate the "collocation points" where the wave functions are handled.

2. They are confined to $N+1$ intervalls.

3. They are continously connected at the "knots" up the the $N-1$-th. derivative.

4. They are positive definite: $u_\alpha^{(N)} > 0$.

5. They are a partition of unity: $\sum_\alpha u_\alpha^{(N)}(x) = 1$.

The feature 1 provides an extremely good description of derivatives up to order N. The feature 2 makes them easy to handle. The feature 4 avoids round-off problems. And the feature 5 provides smooth interpolation.

For many purposes it is advantagous to switch to the so called collocation representation where the wave function is represented on "collocation points" ξ_μ

$$\psi(\xi_\mu) = B_{\mu\alpha}\psi^\alpha \quad \text{where} \quad B_{\mu\alpha} = u_\alpha^{(N)}(\xi_\mu)$$

The coefficients ψ_α then can be replaced by the wave functions $\psi(\xi_\mu)$ because both are equivalent and related by $B_{\mu\alpha}$ or its inverse. The $\psi(\xi_\mu)$ provide a local representation in that the action of a local operator is diagonal, i.e.

$$V(x)\psi(x) \rightarrow V(\xi_\mu)\psi(\xi_\mu)$$

However, the price is that derivatives are represented by fully packed matrices

$$\frac{d}{dx}\psi(x) \rightarrow (D)_\nu{}^\mu \psi(\xi_\mu)$$

On the other hand, once one has full matrices anyway, one does not hesitate to go to any desired order, in particular because the smoothing properties of the Basis Splines avoid the danger of increasing round-off errors. That reduces the expense because than one can obtain a given precision with less grid points. For example we can reduce a mesh of 30 points in a second order finite difference scheme to a mesh of 12 points in a 5-th. order Basis Spline representation and yet obtain higher precision. Thus a computation of the kinetic energy ($\sim d^2/dx^2$) in one dimension requires $4 \star 30$ operations in finite differences versus $12 \star 12$ operations with Basis Splines; that is still slightly advantagous for the finite differences. The situation changes in three dimensions where $4 \star 30^3 = 108000$ looses against $12^4 \sim 20000$. It comes in addition that the matrix multiplication involved with the Basis Splines is easily vectorizable and possibly well suited for parallel machines. Altogether, we think that Basis Splines are a promising tool for large scale computations, particularly in many dimensions.

4 Iterating the Dirac Equation

Another key problem in the numerical simulation of nuclear collective phenomena is the propagation of the wave functions and fields. The time-dependent schemes need a fast algorithm for the time evolution whereas the stationary schemes (as Hartree-Fock , constrained Hartree-Fock and cranking) need a fast static iteration scheme. We will here present and discuss a very fast iteration scheme, the so called damped gradient iteration [8,9]. In particular, we will consider the adaption of these schemes to iterating the Dirac equation which plays a central role in the relativistic mean-field models. A fast solution of the Dirac equation is also needed in other areas of physices wherever strong fields are encountered, as in quantum electrodynamics in strong external fields [7,10] or the bag models of the nucleon [11].

4.1 The damped gradient iteration nonrelativistically

The task is to solve the eigenvalue equation

$$(h - \epsilon_\alpha)\varphi_\alpha = 0 \tag{3}$$

for the lowest A eigenvalues, $\alpha = 1, \ldots, A$. This is done by the iteration

$$\varphi_\alpha^{(n+1)} = \mathcal{O}\{[1 - \mathcal{D}(h - \epsilon_\alpha^{(n)})]\varphi_\alpha^{(n)}\} \tag{4}$$

where the upper index denotes the order of iteration, $\mathcal{O}$ means orthonormalisation, $\epsilon_\alpha^{(n)} =< \varphi_\alpha^{(n)}|h|\varphi_\alpha^{(n)} >$, and $\mathcal{D}$ is a damping functional. The method is called gardient iteration because the $(h - \epsilon_\alpha^{(n)})\varphi_\alpha^{(n)}$ is a gradient with respect to φ_α^* variation in the multidimensional landscape of the energy functional $\mathcal{E} = \sum_\alpha < \varphi_\alpha^{(n)}|h|\varphi_\alpha^{(n)} >$. It is called damped gradient iteration because of the damping functional $\mathcal{D}$ [8].

The efficieny of the iteration (4) depends sensitively on a good choice for $\mathcal{D}$. The simplestmost choice is just a small number $\mathcal{D} = \delta$. This yields the straight gradient iteration which is very slow because $\delta < 2/h_{max}$ is required to ensure convergence towards the lowest eigenvalue (h_{max} is the maximum representable eigenvalue on the given grid). Usually, the kinetic energy determines the upper end of the spectrum, $h_{max} \sim \frac{p_{max}^2}{2m}$, and the high p_{max} slows down the iteration. It is obvious that one should try to include the kinetic energy in the damping $\mathcal{D}$ to remove the most critical piece in h. This leads to the choice

$$\mathcal{D} = \frac{\delta}{E_0 - \frac{p^2}{2m}} \tag{5}$$

where in addition to δ the E_0 is a numerical parameter to be adjusted to give optimal convergence. One usually has to choose E_0 of the order of the variations of the potential in h. The step (4) with the damping (5) constitutes the damped gradient iteration which is very succesful in all sorts of nonrelativistic applications.

4.2 Damped steps for the Dirac equation

We now want to extend the iteration (4) with (5) to the case of the Dirac equation. The Dirac Hamiltonian reads essentially

$$h_D = \vec{\alpha}\cdot\vec{p} + \gamma_0 m + \gamma_0 U_S + U_V = \begin{pmatrix} m + U_S + U_V & \vec{\sigma}\cdot\vec{p} \\ \vec{\sigma}\cdot\vec{p} & -m - U_s + U_V \end{pmatrix} \tag{6}$$

where U_S is a scalar potential and U_V a vector potential. We are interested in the first few positive energy bound states. The problem is that $h_D - \epsilon_\alpha$ is not positive definite due to the negative energy states in the Dirac spectrum. A further problem is that we have now *two* high energies in the spectrum, first the $h_{max} \sim \sqrt{m^2 + p_{max}^2}$ as before, and second the particle-antiparticle gap $2m$. Both these high energies can slow down the iteration and a possible iteration scheme needs to cope with both hindrances. We have investigated three alternative iteration schemes which all solve these problems more or less efficiently.

In the *effective Schrödinger step* we transform the Dirac equation into an effective Schrödinger equation for the upper wave function component alone. This yields an effective Hamiltonian

$$h_{eff} = m + U_S + U_V + \vec{\sigma} \cdot \vec{p} \frac{1}{m + \epsilon_\alpha + U_S - U_V} \vec{\sigma} \cdot \vec{p} \tag{7}$$

The effective Schrödinger step $\mathcal{S}_{eff}$ is then just the nonrelativistic step (4) with the h_{eff} inserted. The fact that h_{eff} themselves depends on the energy ϵ_α causes no real problem because we iterate it anyway. We expect that the step $\mathcal{S}_{eff}$ works well in nearly nonrelativistic cases where the potential $U_S + U_V$ is weak.

In the *quadratic Dirac step* we first construct a positive definite operator by considering $(h_D - e_{shift})^2$ where e_{shift} is some energy shift to be placed roughly near the desired eigenvalue. An obvious damping functional then has the structure of the squared Dirac kinetic energy. Thus

$$\mathcal{S}_{quad} = 1 - \delta(\vec{\alpha} \cdot \vec{p} + \gamma_0 M_0 - E_0)^{-2}[(h_D - e_{shift})^2 - < (h_D - e_{shift})^2 >_\alpha] \tag{8}$$

where the expectation value with $\varphi_\alpha^{(n)}$ is subtracted to project out the desired eigenvalue. Note that the Dirac damping has two numerical parameters, M_0 and E_0. That is very welcome because we need to account for two high energy parts in the spectrum (h_{max} and $2m$). However, the quadratic step first squares the problem before solving it. One has to see how it performs.

In the *linear Dirac step* we try to use h_D directly and the Dirac kinetic energy as damping functional. Thus

$$\mathcal{S}_{lin} = 1 - \delta(\vec{\alpha} \cdot \vec{p} + \gamma_0 M_0 - E_0)^{-1}[h_D - < h_D >_\alpha] \tag{9}$$

This step is the most promising if it is possible to find damping parameters M_0 and E_0 such that $\mathcal{D}[h_D - \epsilon_\alpha]$ is positive definite.

We have tested these three steps under various conditions, for spherical one dimensional calculations as well as for full three dimensional ones, and for finite difference schemes as well as for Basis Spline representations. We show here the results for the three dimensional calculations using Basis Splines. Four different physical cases have been studied, all employing Gaussian-type potentials $U(r) = U(0)\exp(-(\frac{r}{\delta r})^2)$:

- Nuclear field: $U_S(0) = -0.60m$, $U_V(0) = +0.45m$ and $\delta r = 10/m$
- Vector field: $U_S(0) = 0m$, $U_V(0) = -2m$ and $\delta r = 1.3/m$
- Scalar field: $U_S(0) = -1.1m$, $U_V(0) = 0m$ and $\delta r = 10/m$
- Strong Scalar field: $U_S(0) = -1.3m$, $U_V(0) = 0m$ and $\delta r = 10/m$

The *Nuclear field* is typical for the relativistic mean-field model discussed in section 2. The *Vector field* is typically encountered in the quantum electrodynamics of strong fields which occurs in connection with heavy-ion collisions. The *Scalar fields* are typical for bag models of the nucleon. There we have distinguished two cases: the *Scalar field* has a positive energy bound state ϵ_0 just above the antiparticle potential well, whereas the ϵ_0 crosses that line in case of the *Strong Scalar field.*

We have scanned a large variety of damping parameters δ, M_0 and E_0. The speed of iteration grows with growing δ until convergence suddenly breaks down. Thus for given M_0 and E_0 the optimal δ is well defined and fairly easy to find. For the M_0 and E_0 there is a wider range of choices. We will not discuss thoses details here. In fig.5 we show the convergence for the various cases and for well chosen damping parameters for each case. As expected, the quadratic step (8) is the slowest. The linear step (9) is the fastest with an amazing speed of achieving 6 digits precision within about 30-70 iterations (depending on the case, of course). The effective Schrödinger step is competitive in the case of the Nuclear field and is looses a bit in the other cases. It completely fails in case of the Strong Scalar field because there the effective mass in h_{eff} of eq. (7) becomes indefinite. Altogether, we see that all methods converge well and fast. Thus we have quite a flexibility to accomodate varying situations. In general, the linear Dirac step (9) seems to be the most promising tool.

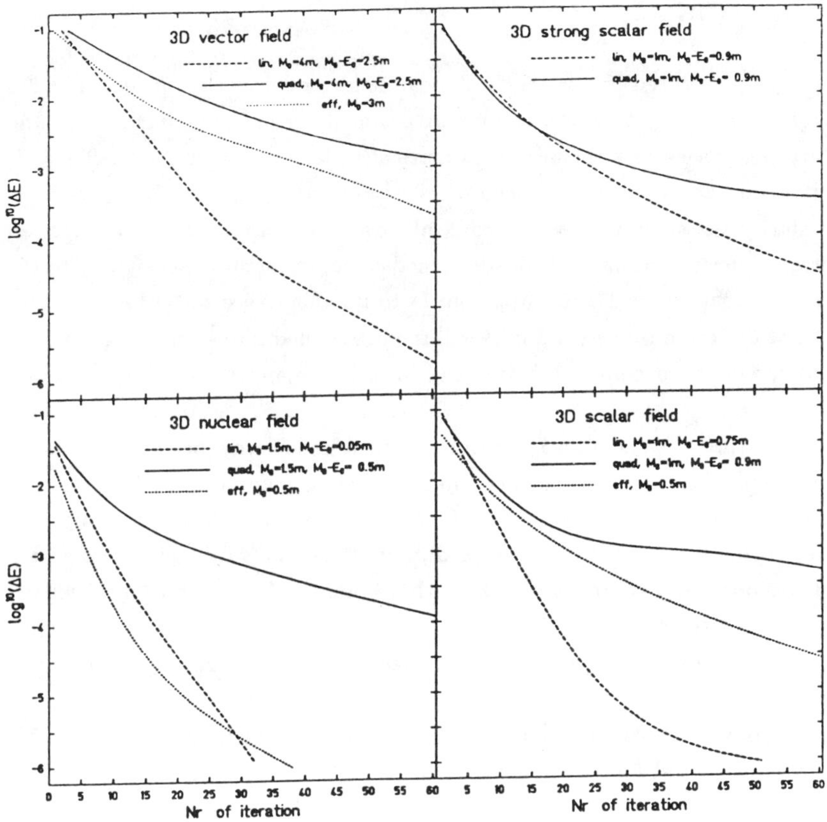

Figure 5:
Convergence of the three steps for the Dirac equation for the four cases studied. The energy fluctuation $\Delta E = \sqrt{< h_D^2 >_\alpha - < h_D >_\alpha^2}$ *is used as a measure of the precision achieved. It is drawn logarithmically versus the number of iteration.*

5 Conclusion

We have presented the relativistic mean-field model of the nucleus and nuclear dynamics. It allows a high precision description of nuclear ground state properties and includes automatically the spin-orbit force. There is some concern about the rather low effective mass in the model. One hope to clarify the situation by accessing new ranges of applications, as e.g. deformed nuclei. And indeed, the deformed calculations show sensitivity to the parametrisation used and in particular to its effective mass. More research in that direction is needed and it will require refined numerical techniques as well as enhanced computing facilities in order to allow sufficient variations of the model.

The time-dependent calculations of medium energy heavy-ion scattering within the relativistic mean-field model show complete spallation of the colliding nuclei, in contrast to nonrelativistic results. This is a relativistic effect coming from the meson retardation. It deserves further investigation for various energy ranges and for other combinations of nuclei. But the computations are extremely demanding and one will have to wait for the new generation of inexpensive vector machines showing up until systematic studies on that topic can continue.

We have also sketched two numerical methods which promise to provide more precise and faster calculations in the future. The one was the spatial representation of wave functions and fields with Basis Splines. Their smooting properties, their stablilty and their good reproduction of derivatives will make them to useful tools for microscopic models of nuclei and nuclear dynamics, in particluar in many dimensional calculations. The other numerical topic has been concerned with the implementation of the damped gradient iteration for solving the stationary Dirac equation. We have managed to get a very fast and reliable iteration scheme which achieves 6 digit precsion within about 50 iteration (depending a bit an the actual case studied). It remains a task for future investigation to develop a similar acceleration for the time step in time-dependent calculations.

References

[1] K. Goeke, P.-G. Reinhard, Proceedings of the Workshop *TDHF and beyond*, Lecture Notes in Physics, Vol. **171** (1982)

[2] P. Quentin, H. Flocard, Ann.Rev.Nucl.Part.Sci. **28** (1978) 523

[3] B.D. Serot, J.D. Walecka, Advances in Nuclear Physics, Vol. **16** (1986)

[4] L.S. Celenza, C.M. Shakin, *Relativistic Nuclear Physics - Theories of Structure and Scattering*, World-Scientific, Singapore

[5] R. Machleidt, K. Holinde, C. Elster, Phys.Rep. *C*149 (1987) 1

[6] P.-G. Reinhard, M. Rufa, J. Maruhn, W. Greiner, J. Friedrich, Z.Phys. **A323** (1986) 13

[7] M. Rufa, P.-G. Reinhard, J. Maruhn, W. Greiner, M.R. Strayer, to appear in Phys.Rev. (1988)

[8] S.-J. Lee et al, Phys.Rev.Lett. **59** (1986) 1171

[9] J. Fink et al, GSI Scientific Report 1987

[10] R.Y. Cusson et al, Phys.Rev.Lett. **55** (1985) 2786

[11] J.J. Bai et al, Z.Phys. **A326** (1987) 269

[12] C. Bottcher, M.R. Strayer, Ann.Phys. **175** (1987) 64

[13] P.-G. Reinhard, R.Y. Cusson, Nucl.Phys. **A378** (1982) 418

[14] R.Y. Cusson, P.-G. Reinhard, M.R. Strayer, J.A. Maruhn, W. Greiner, Z.Phys. **A320** (1985) 475

[15] see contribution of B. Müller in these proceedings

[16] see contribution of L. Wilets in these proceedings

HYDRODYNAMIC MODEL OF HIGH-ENERGY HEAVY-ION COLLISIONS

JOACHIM A. MARUHN
Institut für Theoretische Physik
Universität Frankfurt
D-6000 Frankfurt 11, West Germany
and
Joint Institute for Heavy Ion Research
Holifield Heavy Ion Research Facility
Oak Ridge, Tennessee 37831

Abstract

The hydrodynamic model is one of the theoretical approaches used successfully for the exploration of the reaction mechanism in high-energy heavy-ion collisions. After describing the model equations and numerical methods I discuss the present status of the comparison to experiment, especially the problem of equilibration and the search for the nuclear equation of state.

INTRODUCTION

The field of relativistic heavy ion collisions provides a fascinating new way of studying the behaviour of nuclear matter under extreme conditions. Whereas in traditional low-energy nuclear reactions, with energies typically of the order of up to several MeV per nucleon, nuclear matter is practically incompressible, this should be quite different at the energies of up to 200 GeV per nucleon studied at the new accelerators. One may estimate the speed of sound in nuclear matter from the incompressibility constant $K \approx 210MeV$ as

$$c_s/c = \sqrt{\frac{K}{9m}} \approx 0.16 \quad ,$$

which corresponds to a bombarding energy of about 12 MeV per nucleon.

Another energy threshold is defined by the *Fermi-energy* in the nucleus, which is of the order of 40 MeV. Exceeding this value with the energy of the incoming projectile nucleons implies that collisions between individual nucleons will not be blocked as effectively by the Pauli principle as they are in low-energy collisions. Taking the necessary caution with respect to quoting exact numbers we may thus infer that *in the region between 10 and 100 MeV per nucleon there should be a transition between the low-energy heavy ion collision behaviour, in which nuclear matter is incompressible and the dynamics is dominated by the mean field, to a regime of compression effects with nucleon-nucleon collisions determining the dynamic evolution.*

If we think of the nucleus in terms of a classical liquid drop, then these collisions should be characterized by the propagation of *shock waves* as occur in supersonic motion of fluids or gases. It is, however, not certain *a priori* whether such an analogy applies, since nuclei are relatively small systems and the establishment of local equilibrium is not trivially assured. Estimates for the mean free path of a nucleon in the nuclear medium at these energies yield values of a few femtometers, neither small nor large compared to the size of the nucleus [1].

The best way to proceed in such an uncertain situation is to study various models making quite different assumptions about the reaction mechanism and to then find out which one of them is favored by experiment. In the following I will concentrate on the hydrodynamic model, in which *instantaneous local equilibrium*, *i. e.* a negligible mean free path, is assumed. Results will also be quoted from the *intranuclear cascade model* [2,4,3], which describes the heavy ion collision as an assembly of essentially free nucleons that interact only in binary pointlike collisions with the scattering result given by the free nucleon-nucleon cross section. Let us now proceed to the definition of the hydrodynamic model. A partial list of references to previous work in this field is given in [5], while [6] gives an overview of the historic development. Independent work with different numerical methods was done at Los Alamos [7].

THE MODEL EQUATIONS

We formulate the equations in an arbitrary inertial frame, for which in the calculations usually the laboratory-fixed frame or the center-of-velocity frame, in which the four-velocities of target and projectile are equal in magnitude, are taken. For the notation we follow [8], *i. e.* four-vectors are defined like

$$x^\mu = \{x^0, x^1, x^2, x^3\} = \{ct, \vec{x}\} = \{ct, x^k\} \quad ,$$

with Greek indices running from 0 to 3 and Latin indices from 1 to 3. The inner product is defined via the metric tensor $g_{\mu\nu}$ which is diagonal with diagonal elements

$$g_{00} = 1,\ g_{11} = -1,\ g_{22} = -1,\ g_{33} = -1 \quad ,$$

while the four-gradient is

$$\partial_\mu = \left\{ \frac{1}{c}\frac{\partial}{\partial t}, \vec{\nabla} \right\} \quad .$$

The equations of motion of relativistic hydrodynamics can be obtained from conservation conditions for the four-current

$$\partial_\mu j^\mu = 0 \quad ,$$

and for the energy-momentum tensor

$$\partial_\mu T^{\mu\nu} = 0 \quad .$$

The four-current density j^μ is defined in terms of the fluid density in the rest frame, n, as

$$j^\mu = n u^\mu / c \quad ,$$

where u^μ denotes the four-velocity of the fluid and is related to the usual three-dimensional velocity $\vec{v}$ via

$$u^0 = \gamma \quad , \quad u^i = \gamma v_i / c \quad ,$$

with γ the familiar relativistic factor

$$\gamma = \frac{1}{\sqrt{1 - \vec{v}^2/c^2}} \quad .$$

u^μ is normalized to $u_\mu u^\mu = 1$. For a perfect fluid the energy-momentum tensor is given by [9]

$$T^{\mu\nu} = (e + p) u^\mu u^\nu + p\, g^{\mu\nu} \quad ,$$

where e is the energy density measured in the local rest frame (including the rest mass of the particles), and p the hydrodynamic pressure. To rewrite these equations in a more familiar form, the individual parts of $T^{\mu\nu}$ in the laboratory frame that correspond to classical quantities are defined as

- the energy density

$$E = T^{00} = \gamma^2 (e + p) - p \qquad (1)$$

- the momentum density

$$M_k = cT^{0k} = \gamma^2(e+p)v_k \quad . \tag{2}$$

In the same way, the four-current can be decomposed into the density in the laboratory frame,

$$\rho = j^0/c = \gamma n \tag{3}$$

and the spatial part given by

$$j^k = \rho u^k = \gamma \rho v^k / c \quad .$$

Utilizing these definitions, the equations of motion can be rewritten in three-dimensional notation as

$$\begin{aligned} \partial_t \rho + \nabla \cdot (\rho \vec{v}) &= 0 \\ \partial_t \vec{M} + \nabla \cdot (\vec{v}\vec{M}) &= -\nabla p \\ \partial_t E + \nabla \cdot (\vec{v}E) &= -\nabla \cdot (\vec{v}p) \end{aligned}$$

These equations are identical in form to the nonrelativistic ones, so that many of the numerical methods can be taken over. The big difference, however, is in the application of the equation of state, which is given *in the local rest frame* as

$$p = p(n,e)$$

What this implies, then, is that when the fields ρ, $\vec{M}$, and E, which are all defined in a fixed system of reference such as the laboratory frame, have been propagated to a new time, the rest frame quantities n and e as well as the velocity $\vec{v}$ have to be found *at each point* such as to satisfy the equations 1–3. Physically this means that the local rest frame has to be found for each fluid element.

THE EQUATION OF STATE

One of the most important motivations for the study of these reactions is the hope that the equation of state of nuclear matter, which is practically unknown experimentally, may be pinned down. To do this, however, one first has to find experimental observables that are sensitive to changes in the equation of state.

Typically the equation of state is quoted in terms of the energy per baryon W as a function of density n and entropy density s, and it is assumed that it can be split up into a contribution for cold matter W_c, called "compressional" energy or sometimes simply the equation of state, and a thermal part that is determined from some relatively simple model, W_{th}. The pressure and speed of sound are then given by the usual thermodynamic definitions as

$$p = n^2 \frac{\partial W_c}{\partial n} + n^2 \frac{\partial W_{th}}{\partial n}\bigg|_s \qquad c_c^2 = \frac{\partial p}{\partial e}\bigg|_s \quad .$$

Some examples of equations of state that have been used in practical calculations are

- the *quadratic* equation of stategiven by

$$W_c(n) = \frac{K}{18 n_0^2}(n - n_0)^2$$

  with K the *incompressibility* constant of nuclear matter and n_0 the equilibrium density. This formula simply extends the quadratic behavior of the energy around the equilibrium state to higher densities. It has the deficiency that the speed of sound exceeds the speed of light already at moderate densities.

- the *linearized* equation of state was developed expressly to avoid this shortcoming. It is given by

$$W_c(n) = \frac{K}{18 n n_0}(n - n_0)^2 \quad .$$

- In both of these cases the thermal part is usually approximated by a nonrelativistic Fermi gas, which should be reasonable for low temperatures where the thermal velocities of the particles are small compared to c and there is too little excitation energy to excite the nucleons or to create new particles like pions. For this case the pressure can simply be expressed as

$$p = \frac{\partial W_c}{\partial n} + \tfrac{2}{3}(e - n W_c(n))$$

- more sophisticated calculations employ a Fermi gas including contributions of the excited states of the nucleon, pions, *etc.* This takes the form of a sum over Bose and Fermi integrals, one for each species of particle. In the calculations interpolation tables for the equation of state have to be employed in this case. This complicated thermal energy is still combined with the parametrized compressional part.

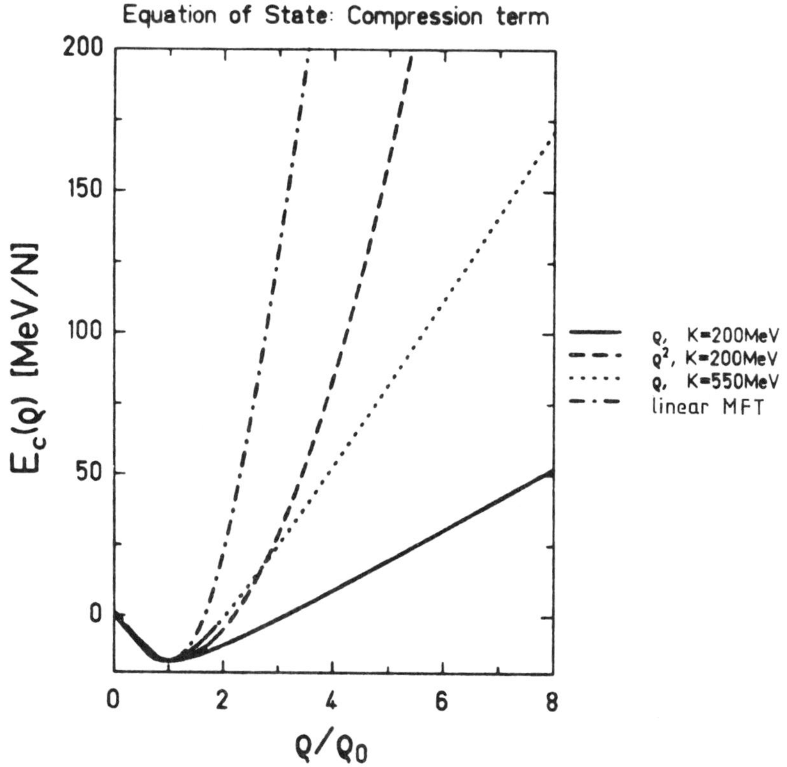

Figure 1: Examples of equations of state used in hydrodynamical model simulations. The curves plotted are for the linear (ρ) and quadratic (ρ^2) equation of state with the indicated incompressibilities, and for a linear mean field model [10].

- many calculations have been done also with more complicated additional effects added, such as the phase transition to a quark gluon gas or a pion condensate.

Figure 1 shows some W_c curves typical of the ones employed in the simulations. Clearly there is immense freedom possible while maintaining the same properties near the ground state of nuclear matter, the only region accessible to conventional experiment. The present purpose of the simulations is largely to narrow the choices in this respect.

OUTLINE OF THE NUMERICAL METHODS

The basic algorithm used for the propagation of the equations of motion is the *SHASTA* algorithm of Boris and Book [11]. Although more accurate algorithms are available nowadays, none of them can match the flexibility of *SHASTA* with respect to equations of state and nothing has to be added to the propagation *per se* to take care of relativity.

Let us briefly review the algorithm. Given a general one-dimensional equation of motion

$$\partial_t f + \partial_x (v_x f) = S(x,t)$$

for a conserved density $f(x,t)$, the algorithm proceeds in the following steps:

- assume that the values f_i, v_i and S_i for the density, velocity, and source are given at grid point number i at the beginning of the timestep. The grid is evenly spaced with a distance of Δx and the current timestep is Δt. Compute:

- dimensionless velocities ε_i and diffusion coefficients ν_i at the midpoints:

$$\begin{aligned} \varepsilon_{i+1/2} &= v_{i+1/2}\frac{\Delta t}{\Delta x} \\ \nu_{i+1/2} &= \tfrac{1}{6}(1-\varepsilon_{i+1/2}^2) \end{aligned}$$

- untransported fluxes f^0 and diffused untransported solution F^D

$$\begin{aligned} f_{i+1/2}^0 &= \nu_{i+1/2}(f_{i+1}-f_i) \\ F_i^D &= f_i + (f_{i+1/2}^0 - f_{i-1/2}^0) \end{aligned}$$

- transport convection factors Q^+ and Q^-

$$\begin{aligned} Q_i^+ &= \frac{\frac{1}{2} - v_i \frac{\Delta t}{\Delta x}}{1 + \frac{\Delta t}{\Delta x}(v_{i+1} - v_i} \\ Q_{i+1}^- &= 1 - Q_i^+ \end{aligned}$$

- transported diffused solution F^{TD}

$$F_i^{TD} = 4\,Q_i^{+2} f_{i+1/2}^0 - 4\,Q_i^{-2} f_{i-1/2}^0 + (Q_i^+ + Q_i^-) f_i + S\Delta t$$

- compute some intermediate fluxes δF^{TD} and f^T

$$\begin{aligned} \delta F_{i+1/2}^{TD} &= F_{i+1}^{TD} - F_i^{TD} \\ f_{i+1/2}^T &= f_{i+1/2}^0 + \nu_{i+1/2}(F_{i+1}^{TD} - F_i^D - F_i^{TD} + F_i^D) \end{aligned}$$

- finally, corrected fluxes f^C and antidiffusion

$$\begin{aligned} f_{i+1/2}^C &= \mathrm{sgn}(f_{i+1/2}^T) \qquad (4) \\ & \quad \max[0, \min(\mathrm{sgn}(f_{i+1/2}^T \delta F_{i-1/2}^{TD}, \\ & \quad |f_{i+1/2}^T|, \mathrm{sgn}(f_{i+1/2}^T)\delta F_{i+3/2}^{TD})] \\ \tilde{f}_i &= F_i^{TD} - (f_{i+1/2}^C - f_{i-1/2}^C) \quad , \qquad (5) \end{aligned}$$

yielding the values $\tilde{f}$ for the end of the time step.

An interesting feature of this method is the relatively complicated maximum-minimum formula used to evaluate the corrected fluxes. This can pose a problem for the vectorizing compilers; in practice I found that the IBM compiler could vectorize these functions only after defining staement functions of the type

$$MAX(A,B) = 0.5 * (A + B + ABS(A - B)) \quad ,$$

while the CRAY compiler had no trouble at all.

The other major part of the numerical problem is the solution of the equations (1-3) to determine the rest frame quantities n and e as well as the velocities $\vec{v}$ from the fields ρ, $\vec{M}$, and E given in the fixed reference frame. We coined the term "untangling" for this process. Since it has to be repeated for every timestep at every point in the grid, an effective algorithm is of paramount importance.

Practically the calculations are performed on a grid of 33×64^2 points with a physical grid spacing of 0.25 fm. The frame of reference is the *center-of-velocity frame*, in which both nuclei have equal velocity and γ-factors. In the beam direction the grid is shrunk by a factor of γ in order to account for the relativistic contraction of both nuclei. Typical computation times are several ten minutes for one collision. The time per timestep is not constant because the code monitors where the grid is empty and does not perform the propagation on those points. Memory requirement is about 20 MB.

RESULTS

As has become apparent from the preceding discussion of the physical background, the theoretical situation in high energy heavy ion collisions is quite open and does as yet not call for high-accuracy calculations: we have a model whose physical validity yet has to be established completely, largely unknown equation of state and dissipative terms, and also the problem of finding experimental observables that provide information on all these unknowns. Clearly it is a time for exploratory calculations and the present state of these will now be reported.

Figure 2 shows as a typical result contour lines for a collision of C+Au at 84 MeV per nucleon and at various impact parameters. At all impact parameters upon contact a shock wave is formed that leads to a heated and compressed zone whose geometric shape reminds one of a "Mach cone". In the final state this geometry is still visible in the residual excited nuclear matter, which will evaporate nucleons and still show some preferred angles of emission because of the remaining collective motion. At larger impact parameters there is also a projectile residue which is deflected to some finite angle (the so-called "bounce-off effect").

This effect has been crucial for getting a first idea about the reaction mechanism. Fig. 3 compares the final states in an intranuclear cascade simulation [4] to that of the hydrodynamic model. While in both the same general division into target residue, projectile residue, and "participants", *i. e.* the highly excited matter left over from the initial impact zone, is visible, the projectile in the cascade calculation is deflected to much smaller angles than in hydrodynamics. Support for the hydrodynamic behaviour came from the experimental observation of *flow*.

The flow tensor is defined from the momenta $\vec{p}^{\alpha}$ of all observed particles as

$$T_{ij} = \sum_{\alpha} \frac{p_i^{\alpha} p_j^{\alpha}}{2m_{\alpha}} \quad .$$

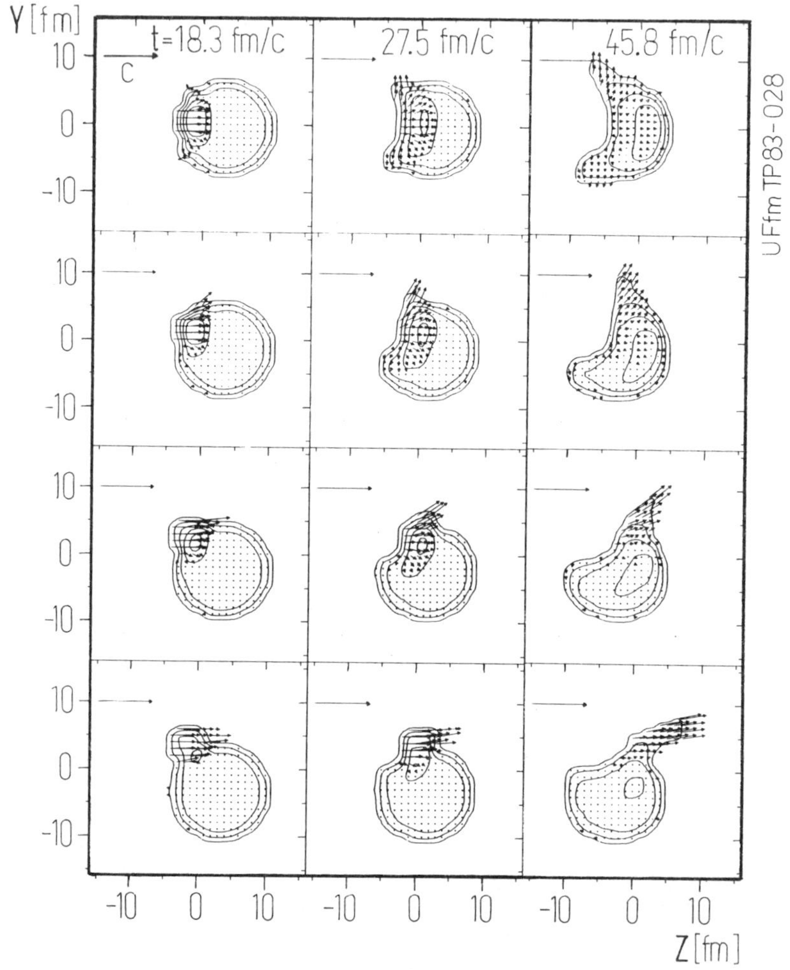

Figure 2: Density contours in a hydrodynamic simulation of C+Au at 84 MeV per nucleon. Time increases from left to right, and impact parameter from top to bottom.

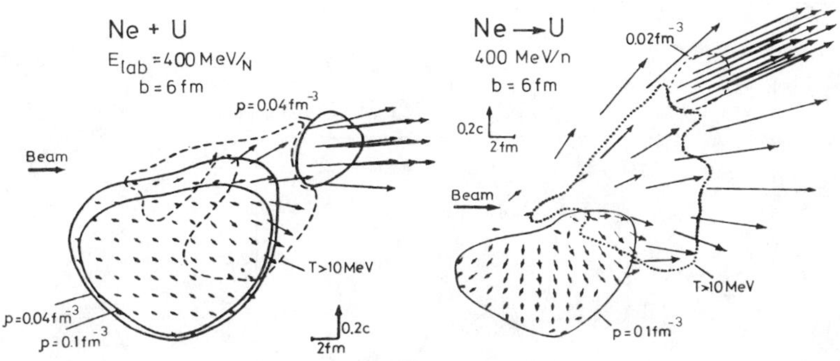

Figure 3: The final state in a collision of Ne+U at 400 MeV per nucleon at 6 fm impact parameter in the hydrodynamic model (right) and an intranuclear cascade [4] (left). The flow is indicated by arrows and regions of high density or high temperature are delineated by isolines.

T defines an ellipsoid in space which indicates the average flow of particles in the final state. It can be characterized in terms of the principal axes and the their orientations. The most important quantity to be examined is the *flow angle*, which is the angle between the axis of largest length and the beam direction and thus describes the sideward flow of matter. Experimentally it is found that for heavier systems there is a systematic shift of flow angle with increasing multiplicity (see figure 4); this effect is not present in the simple cascade calculation but can be qualitatively reproduced by hydrodynamics if, as is usual, it is assumed that increasing multiplicity means decreasing impact parameter.

The conclusion to be drawn is thus that at least some equilibration is taking place and that the hydrodynamic approach is closer to the facts than the independent particle cascade model — although one should not conclude that local equilibrium is exactly fulfilled. It was then found that the crucial ingredient lacking in these older cascade models was a mean field. Introducing that in the framework of the so-called Vlasov-Uehling-Uhlenbeck model [14] led to the presence of flow also in the microscopic calculation.

After getting this assurance of some equilibration, it starts making sense to talk about determining the equation of state from the experiments, which again raises the problem of observables. An early attempt [15] to derive it

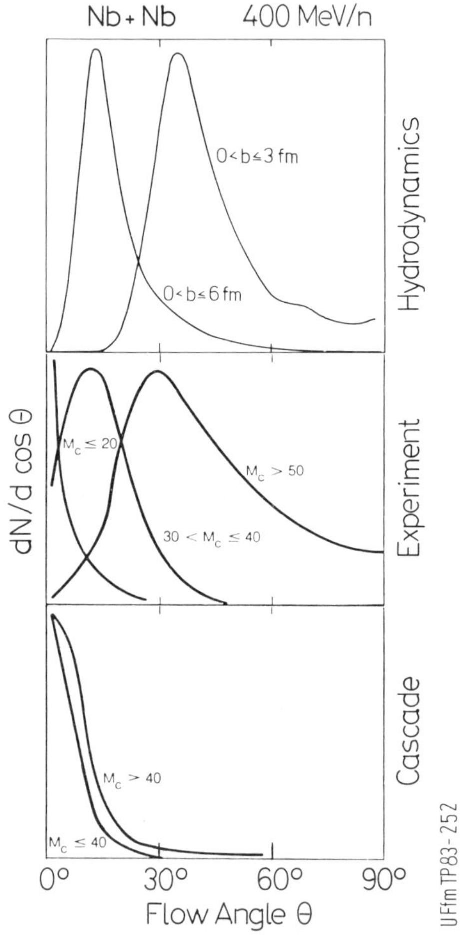

Figure 4: Distribution of flow angles in final states of a Nb+Nb reaction at 400 MeV per nucleon [12,13]. The events are plotted in multiplicity (M_c) groups, showing a systematic shift to higher flow angles for more central collisions, which is not present in the cascade. The hydrodynamic model shows such a shift as a function of impact parameter.

from the pion yields was shown to lead to ambiguous results [16], even without considering all the other uncertainties associated with pions in nuclear matter. More recently also the inclusion of viscosity in the hydrodynamic model has raised new questions [18]. Viscosity can be introduced *in the non-relativistic formulation* by replacing the pressure by the Maxwellian stress tensor

$$\begin{aligned}\sigma_{ij} &= -p\delta_{ij} \\ &\quad +\eta\left(\frac{\partial v_i}{\partial x_j}+\frac{\partial v_j}{\partial x_i}-\tfrac{2}{3}\delta_{ij}\nabla\cdot\vec{v}\right) \\ &\quad +\zeta\delta_{ij}\nabla\cdot\vec{v} \quad .\end{aligned}$$

In practice the shear viscosity η seemed to be more interesting than the bulk viscosity ζ, and the order of magnitude was estimated as [17]

$$\eta \approx 30 - 60\ MeV/fm^2c = 0.16 - 0.32\ TP$$

Figure 5 shows the influence of both the equation of state and the vis-

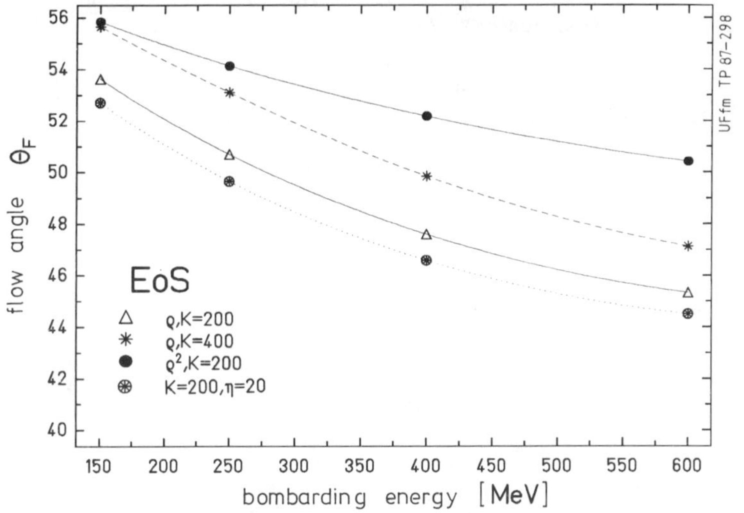

Figure 5: Dependence of the flow angle excitation function on the equation of state and the viscosity in Au+Au. The curves show results for a linear ($\rho, K = 200$ and $K = 400$), quadratic (ρ^2), and linear with viscosity ($K = 200MeV, \eta = 20MeV/fm^2c$) equations of state.

cosity on the flow angle; it is quite disappointing and certainly does not

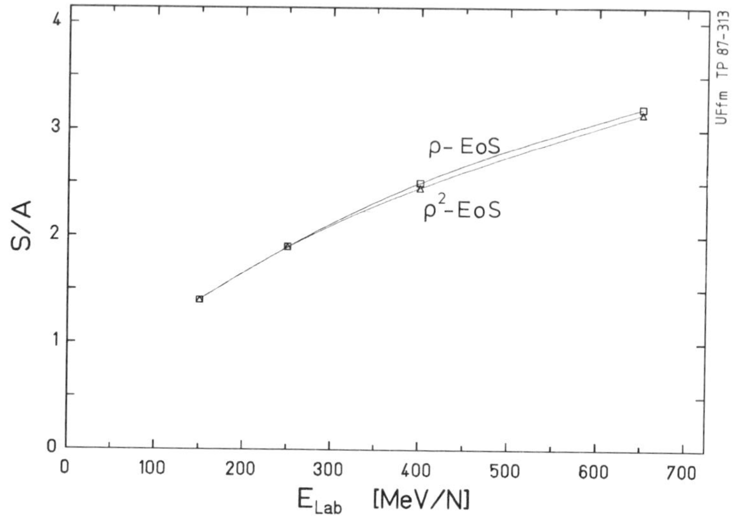

Figure 6: Dependence of the excitation function of the entropy per nucleon in Au+Au on the equation of state.

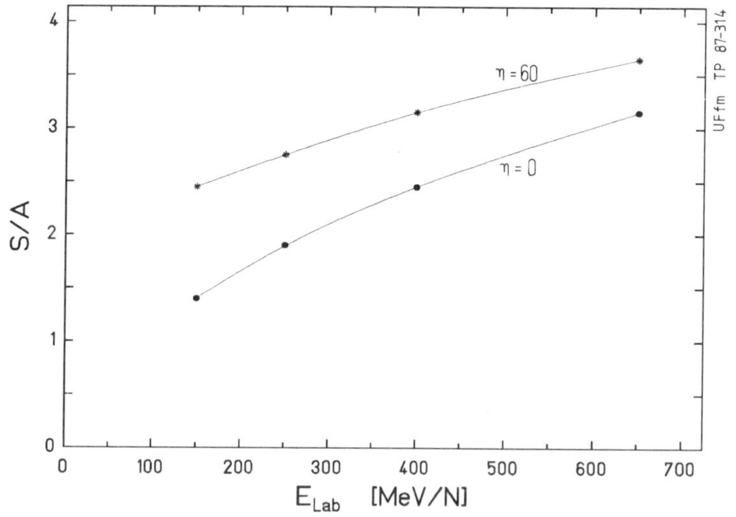

Figure 7: Dependence of the excitation function of the entropy per nucleon in Au+Au on the viscosity coefficient. The diamonds indicate data processed from experiment [20] as in [19].

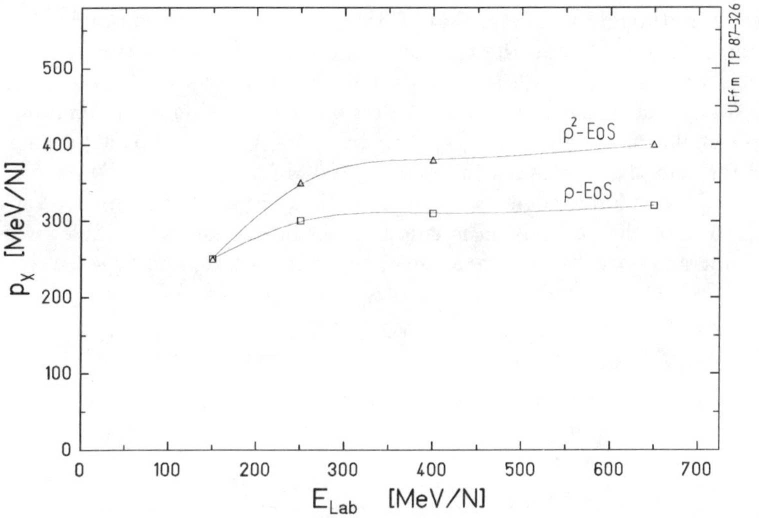

Figure 8: Dependence of the excitation function of the transverse momentum in Au+Au on the equation of state.

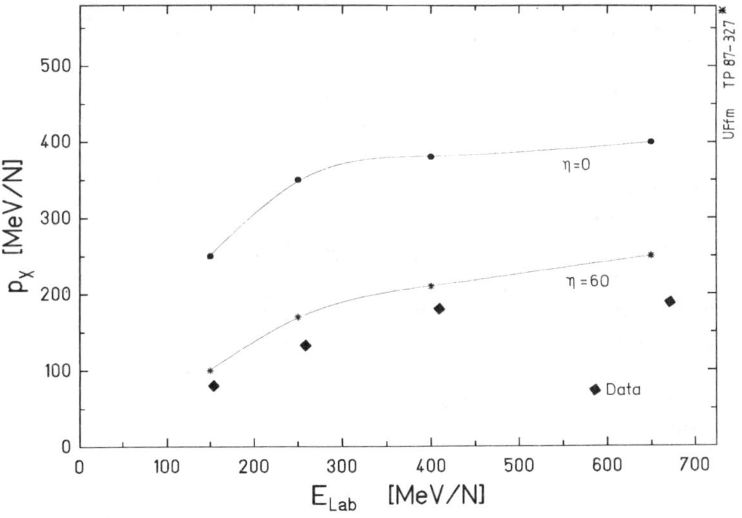

Figure 9: Dependence of the excitation function of the transverse momentum in Au+Au on the viscosity coefficient. Experimental data from [21].

correspond to anything of measurable size. One thus has to look at other observables. Two rather more interesting ones were found: the entropy (which can, though in a somewhat indirect way, be related to the deuteron-to-proton ratio [19]), and the transverse momenta in the reaction plane, p_X [21,22]. Figures 6 through 9 show that the dependence on the viscosity in both cases is more pronounced than that on the equation of state. In all of these calculations the colliding system was Au+Au, the impact parameter $b = 3$ fm, and the incompressibility $K = 700$ MeV.

The present knowledge about the sensitivity of the observables on both ingredients of the calculation is summarized in the following table, where "-" denotes practically no dependence, "+" a moderate, and "++" a strong one:

Observable	Equation of state	Viscosity
Flow angle	+	+
Flow ratio	+	+
Entropy	−	++
Transverse momentum	−	+
Rapidity distribution	−	+

I have also added quantities I have not discussed here for completeness.

BENCHMARK RESULTS

It turned out to be quite interesting to compare the performance of various computers with respect to the hydrodynamical problems discussed above. I tried to run a simple two-dimensional problem on all the computers readily available to me. It had to be a slightly simplified version because the code had to be translated for some of the tests.

Desktop *PC*-s are not really up to number-crunching yet, but are certainly useful during the development stage because of the exceedingly user-friendly software available. Getting the code to run using Microsoft Fortran or C was very easy (There was no noticeable speed difference between these languages). However, the running times of the finished product are not good enough for production-type calculations. Of course the "old" PC-XT is completely disqualified.

This cheap machine can get a tremendous boost, however, by inserting a monoputer board, which brings its numbercrunching capability way beyond the AT-class computers. I was able to test the code on a T800-monoputer board, in both a C and and Occam 2 version (a Fortran compiler is also available). There still seemed to be a few bugs in the C-system, and Occam

Table 1: Results of the benchmark test for a grid of 100 × 100 points, and additional criteria for system choice.

Computing system	Language	$t[s]^a$	$t/t_0{}^b$
PC-Clone[c]	Fortran, C	135	321
IBM PS/2-50	Fortran, C	44	104
Transputer[d]	C, Occam 2	9	21
VAX 785	Fortran	5	12
IBM 3080-4E	Fortran	0.42	1.0
IBM Vector[e]	Fortran	0.23	0.54
CRAY 2	Fortran	0.05	0.12

[a]CPU-time per timestep
[b]relative to IBM 3080-4E
[c]8088/8087 4.77MHz
[d]PC-Clone with INMOS Monoputer T800, 2MB
[e]IBM 3080-4E with Vector Facility

is quite an inconvenient language for scientific applications (note that again speed is about the same for both languages), but the results of the test are certainly interesting, making this the lowest-cost number-crunching solution included in the tests. I look forward to using the quadputer board as well.

The IBM mainframe tests were done at the GSI computing center. The vector facility installed there was also used, but gave a relatively disappointing acceleration of only about a factor 2. This is probably due to the small fast memory of that device, which requires judicious programming for optimum speed to be obtained. Since there was not enough time to do that, I stopped after having made the program fully vectorizable by the compiler. For the CRAY 2 the compiler also indicated full vectorization. Clearly for very large problems there is no alternative to mainframe computers, but there are many problems for which a desktop system such as on based on transputers could provide a very cost-effective alternative.

FUTURE DIRECTIONS

We have seen that there is still a long way to go before firm conclusions can be drawn concerning the equation of state and dissipative terms. What is needed is a systematic study of both the hydrodynamic and microscopic models: only if both yield comparable agreement with experiments for similar material properties can we be convinced we have obtained a correct description. Also it is clear that no single observable provides sufficient information; again as many observables as possible will have to be studied simultaneously. Such a project is underway and will certainly require a huge amount of computing, for which Gigaflop-type processors are certainly desirable. This would also allow for higher accuracy in hydrodynamics and the study of rare events in the microscopic models.

ACKNOWLEDGEMENTS

The author wishes to express his gratitude to all who in many years of collaboration contributed to this project, the principal ones being W. Greiner, H. Stöcker, G. Graebner, G. Buchwald, B. Waldhauser, J. Theis, T. Rentzsch, and W. Schmidt.

The Joint Institute for Heavy Ion Research has as member institutions the University of Tennessee, Vanderbilt University, and the Oak Ridge National Laboratory; it is supported by the members and by the Department of Energy through Contract Number DE-AS05-76ERO-4936 with the University of Tennessee.

REFERENCES

[1] M. T. Collins and J. J. Griffin, Nucl. Phys. **A348**, 63 (1980)

[2] J. Cugnon, T. Mizutani, and J. Vandermeulen, Nucl. Phys. **A352**, 505 (1981).

[3] Y. Yariv and Z. Fraenkel, Phys. Rev. **C20**, 2227 (1979).

[4] K. K. Gudima and V. D. Toneev, Yad. Fiz. **27**, 658 (1978); **31**, 1455 (1980).

[5] H. Stöcker, J.A. Maruhn, and W. Greiner, Z. Physik **A290**, 297 (1979); H. Stöcker, J.A. Maruhn, and W. Greiner, Z. Physik **A293**, 173 (1979); H. Stöcker, J.A. Maruhn, and W. Greiner, Phys. Rev. Lett. **44**, 725 (1980); H. Stöcker, C. Riedel, Y. Yariv, L.P. Csernai, G. Buchwald, G. Graebner, J. Maruhn, W. Greiner, K. Frankel, M. Gyulassy, B. Schürmann, G. Westfall, J.D. Stevenson, J.R. Nix, and D. Strottmann, Phys. Rev. Lett. **47**, 1807 (1981); H. Stöcker, G. Buchwald, L.P. Cser-

nai, G. Graebner, J.A. Maruhn and W. Greiner, Nucl. Phys. **A387**, 205c (1982).

[6] W. Scheid, R. Ligensa, and W. Greiner, Phys. Rev. Lett. **21**, 1479 (1968); W. Scheid and W. Greiner, Z. Physik **226**, 365 (1969); G. F. Chapline, M. H. Johnson, E. Teller, and M. S. Weiss, Phys. Rev. **D8**, 4302 (1973); C. Y. Wong and T. A. Welton, Phys. Lett. **49B**, 243 (1974); W. Scheid, H. Müller, and W. Greiner, Phys. Rev. Lett. **32**, 741 (1974); M. I. Sobel, P. J. Siemens, J. P. Bondorf, and H. A. Bethe, Nucl. Phys. **A251**, 502 (1972); J. Hofmann, H. Stöcker, U. Heinz, W. Scheid, and W. Greiner, Phys. Rev. Lett. **36**, 88 (1976).

[7] A. A. Amsden, G. F. Bertsch, F. H. Harlow, and J. R. Nix, Phys. Rev. Lett. **35**,905 (1975); A. A. Amsden, A. S. Goldhaber, F. H. Harlow, and J. R. Nix, Phys. Rev. **C17**, 2080 (1978); J. R. Nix, D. Strottmann, Y. Yariv, and Z. Frankel, Phys. Rev. **C25**, 2491 (1982).

[8] C. W. Misner, K. S. Thorne, and J. A. Wheeler, "Gravitation" (W. H. Freeman and Co., San Francisco 1973).

[9] L. D. Landau and E. M. Lifshitz, "Fluid Mechanics" (Pergamon Press, London 1959).

[10] J. D. Walecka, Ann. Phys. **83**, 491 (1974); B. Serot and J. D. Walecka, Advances in Nuclear Physics (Plenum 1985).

[11] J. P. Boris and D. L. Book, J. Comp. Phys. **11**, 38 (1973).

[12] H. A. Gustafsson *et al.*, Phys. Rev. Lett. **52**, 1590 (1984).

[13] G. Buchwald, G. Graebner, J. Theis, J.A. Maruhn, W. Greiner, and H. Stöcker, Phys. Rev. Lett. **52**, 1594 (1984)

[14] G. F. Bertsch, H. Kruse, and S. das Gupta, Phys. Rev. **C29**, 673 (1984); H. Kruse, B. V. Jacak, and H. Stöcker, Phys. Rev. Lett. **54**, 289 (1985); J. J. Molitoris and H. Stöcker, Phys. Rev. **C32**, 346 (1985).

[15] J. W. Harris *et al.*, Phys. Lett. **153B**, 377 (1985); R. Stock, Phys. Rep. **135**, 259 (1986); R. Stock *et al.*, Phys. Rev. Lett. **49**, 1236 (1982).

[16] J.A. Maruhn and H. Stöcker, Z. Physik **A327**, 75 (1987).

[17] P. Danielewicz, Phys. Lett. **B146**, 141 (1984); B. Schürmann, to be published.

[18] T. Rentzsch, W. Schmidt, J. A. Maruhn, H. Stöcker, and W. Greiner, to be published.

[19] D. Hahn and H. Stöcker, LBL-preprints no. 21386 and 22378 (1986).

[20] K. G. R. Doss *et al.*, submitted to Phys. Rev. C.

[21] K. H. Kampert, Dissertation, University of Münster, 1986.

[22] P. Danielewicz and G. Odyniec, Phys. Lett. **175**, 146 (1985).

SOLITON BAG MODELS

Lawrence Wilets*
Institute for Nuclear Theory, Department of Physics, FM-15
University of Washington, Seattle WA 98195, U.S.A.

Abstract Soliton models are well-suited for dynamical calculations, such as hadron-hadron interactions and collisions, since for each variable in the the Lagrangian the time derivative of that variable also appears. For such models, constrained (deformed) mean field solutions provide a basis for generator coordinate dynamical calculations. This requires the solution of a large number of coupled, nonlinear, differential equations involving the quark and scalar fields. The Henyey-Wilets method reduces the problem to the solution of a set of coupled, linear, inhomogeneous, differential equations to be iterated. In the chromodielectric model, color confinement is effected by the self and mutual interactions of the quarks through the chromoelectric field. This requires the self-consistent calculation of the gluon propagator in a spatially varying dielectric function. This now involves the solution of a set of coupled, nonlinear integro-differential equations, which can be linearized and solved by iteration. The problem is computation intensive.

INTRODUCTION

Numerous physical problems fall into the category of coupled nonlinear differential equations subject to boundary conditions equal in number to the number of first order equations. I will begin by stating a concrete example, namely the Friedberg-Lee nontopological

* On sabbatical leave 1987-88 to LBL, Berkeley, CA, and SLAC, Stanford, CA.

model in the mean field approximation. Then I will describe numerical methods for solving this problem. This will be followed by a description of related, but more complex problems, some involving nonlinear integro-differential equations.

Soliton models are well-suited for dynamical calculations since for each variable in the Lagrangian its time derivative also appears. This is not true of MIT-type models, Thus a Hamiltonian can be constructed and powerful techniques available from nuclear many body theory, especially the generator coordinate method, can be brought to bear.

THE NONTOPOLOGICAL SOLITON MODEL

The Friedberg-Lee non-topological soliton model[1-4] is described by a covariant, gauge-invariant Lagrangian density

$$\mathcal{L} = \mathcal{L}_q + \mathcal{L}_\sigma + \mathcal{L}_{q,\sigma} + \mathcal{L}_G \,, \tag{1}$$

where the various components have the following meaning:

The Dirac term is

$$\mathcal{L}_q = \overline{\psi}(i\gamma^\mu D_\mu - m)\psi \,, \tag{2}$$

where m is the diagonal flavor matrix for current masses. For the light, u and d, quarks, the current mass is very small, usually taken to be zero. The field operator ψ is taken to incorporate flavor and color as well as the Dirac indices, and D_μ is the gauge-covariant derivative

$$D_\mu = \partial_\mu + i\tfrac{1}{2} g_s \lambda_c A^c_\mu \,. \tag{3}$$

The λ^c are the eight Gell-Mann color matrices.

The scalar field term is

$$\mathcal{L}_\sigma = \tfrac{1}{2}\partial_\mu\sigma\,\partial^\mu\sigma - U(\sigma) \,. \tag{4}$$

The scalar field σ represents the gluon condensate.

The fermion-scalar interaction term is

$$\mathcal{L}_{q,\sigma} = -g(\sigma)\overline{\psi}\psi \,. \tag{5}$$

This term explicitly breaks chiral invariance since $g(\sigma)$ is like an effective mass. Various forms for $g(\sigma)$ have been considered. The original Friedberg-Lee form is $g_0\sigma$. In fact, a chiral version of the model[5] has no such coupling in the model Lagrangian, but an effective coupling does arise from the gluonic self-interaction of the quarks.

The gluonic part of the Lagrangian is

$$\mathcal{L}_G = -\tfrac{1}{4}\kappa(\sigma)F^c_{\mu\nu}F^{\mu\nu}_c\,. \tag{6}$$

The gauge field tensor is given by

$$F^a_{\mu\nu} = \partial_\mu A^a_\nu - \partial_\nu A^a_\mu + g_S f^a{}_{bc}A^b_\mu A^c_\nu\,, \tag{7}$$

where the $f_{abc} = f^a{}_{bc}$, etc. are the SU(3) structure factors. In the Abelian (one gluon exchange) approximation, to which we restrict discussion here, the nonlinear terms in (7) are absent and the resulting gluonic equations are the same as Maxwell's equations except for the color matrices.

The sigma-field self interaction is taken to be of the quartic form

$$U(\sigma) = \frac{a}{2}\sigma^2 + \frac{b}{3!}\sigma^3 + \frac{c}{4!}\sigma^4 + B\,. \tag{8}$$

The constants a, b and c are fixed within a range so that $U(\sigma)$ has two minima, one at $\sigma = 0$ and a lower minimum at $\sigma = \sigma_v$, the vacuum value. Then $U'(0) = U'(\sigma_v) = 0$. The constant $B \equiv U(0)$ is determined to make $U(\sigma_v) = 0$ and has the same meaning as the MIT bag constant, namely the energy density of deconfinement, or bag pressure. The polynomial form for $U(\sigma)$ is limited to fourth order for simplicity and renormalizability, although the latter is not compelling.

The dielectric function $\kappa(\sigma)$ satisfies the following conditions in order to guarantee absolute color confinement:

$$\kappa(0) = 1, \qquad \kappa(\sigma_v) = 0. \tag{9 a, b}$$

In addition, the condition

$$\kappa'(\sigma_v) = 0 \tag{9 c}$$

must be imposed in order to assure regular behavior for σ as $\sigma \to \sigma_v$. The particular form of $\kappa(\sigma)$ is not crucial so long as the above properties are satisfied. One particular choice[6] has the further properties

$\kappa(\sigma < 0) = 1$ and $\kappa'(0) = \kappa''(0) = 0$. Since $\kappa'(\sigma)$ and $\kappa''(\sigma)$ appear in the equation for the gluon propagator, these derivatives were made continuous at $\sigma = 0$. The form used in ref. (6) is

$$\kappa(\sigma) = 1 + \theta(\sigma)x^3(3x - 4), \tag{10}$$

with $x = \sigma/\sigma_v$; it can be generalized to

$$\kappa(\sigma) = 1 + \theta(\sigma)x^n(nx - (n + 1)). \tag{11}$$

A restriction to $n > 2$ assures continuity of the second derivative at $\sigma = 0$, but even this is not required for cases where σ does not go through zero.

THE MEAN FIELD APPROXIMATION

Because the model contains a complete Lagrangian, the system admits to dynamical calculations. As a preliminary to dynamical calculations, however, we consider first static solutions to the field equations. The simplest of these is the mean field approximation (MFA). We first neglect gluonic interactions and restrict consideration to valence quarks (no vacuum distortion at this stage). The Hamiltonian can then be written

$$H = \int d^3r\, \mathcal{H}(\vec{r}),$$

$$\begin{aligned}\mathcal{H}(\vec{r}) = \psi^\dagger(\vec{r})\big[-i\,\vec{\alpha}\cdot\vec{\nabla} + \beta g(\sigma(\vec{r}))\big]\psi(\vec{r}) \\ +\tfrac{1}{2}\pi(\vec{r})^2 + \tfrac{1}{2}|\vec{\nabla}\sigma(\vec{r})|^2 + U(\sigma),\end{aligned} \tag{12}$$

where $\pi = \dot{\sigma}$.

In the MFA, we treat $\sigma(\vec{r})$ as a c-number and consider only valence quarks. Extremization of the expectation value of H with respect to valence quark functions $\psi_k^\dagger$ (subject to the normalization constraint) and with respect to σ leads to the coupled set of equations

$$(\vec{\alpha}\cdot\vec{p} + \beta g(\sigma))\psi_k = \epsilon_k\psi_k\,, \tag{13a}$$

$$-\nabla^2\sigma + U'(\sigma) + g'(\sigma)\sum_{k(valence)}\psi_k^\dagger\beta\psi_k = 0. \tag{13b}$$

The first is a linear eigenvalue problem (if σ is given); the second is a non-linear inhomogeneous equation (if the ψ_k are given). Satisfaction of the extremization conditions also eliminates terms linear in $\sigma - \sigma(\mathrm{MFA})$ in the expectation value of the energy. In addition to employing the mean field approximation, the set of equations (13) neglects the negative energy sea of quarks, i.e. vacuum distortion.

In the special case that $\sigma(\vec{r})$ is spherically symmetric, the Dirac equation functions can be written

$$\psi_n = \begin{pmatrix} u_n(r) \\ i\vec{\sigma}\cdot\hat{r}\; v_k(r) \end{pmatrix} \mathcal{Y}_{\kappa m} \,, \tag{14}$$

where $\mathcal{Y}_{\kappa m} \equiv \mathcal{Y}^{\ell}_{jm}$ is the two-component Pauli spinor harmonic,

$$\mathcal{Y}^{\ell}_{jm} = \sum_{m_\ell m_s} <\ell\, m_\ell\, \tfrac{1}{2}\, m_s | jm>\, Y_{\ell m_\ell}\, \chi_{m_s} \,, \tag{15}$$

and n encompasses κ, m and a radial quantum number. The Dirac quantum number

$$\kappa = (j + \tfrac{1}{2})\, (-1)^{j-\ell+\frac{1}{2}} \tag{16}$$

denotes both j and ℓ. In this case, Eqs. (13) can be replaced by

$$\left(\frac{d}{dr} + \frac{(\kappa+1)}{r}\right) u_n + (g(\sigma) + \epsilon_n)\; v_n = 0, \tag{17a}$$

$$\left(\frac{d}{dr} - \frac{(\kappa-1)}{r}\right) v_n + (g(\sigma) - \epsilon_n)\, u_n = 0, \tag{17b}$$

$$-\frac{1}{r}\frac{d^2}{dr^2}\, r\sigma + U'(\sigma) + \frac{g'(\sigma)}{4\pi} \sum_{n(valence)} (u_n^2 - v_n^2) = 0 \,, \tag{17c}$$

with the normalization condition

$$\int_0^\infty r^2 dr\, (u_k^2 + v_k^2) = 1. \tag{17d}$$

We have assumed here that the quark distribution is also spherically symmetric. The generalization to non-spherical systems has also been effected[7].

We now consider various techniques for solving Eqs. (17) or their non-spherical generalizations.

NUMERICAL TECHNIQUES

Eqs. (17 a, b) represent a linear eigenvalue problem, for which numerous methods are available (c.f. refs. 4, 8). Eq. (17 c) is a second order nonlinear differential equation coupled to the solutions of (17 a, b, d).

The standard methods for solving nonlinear differential equations are variations of the well-known Newton-Raphson method for non-linear algebraic equations. This consists of linearization about an approximate solution:

Consider first the algebraic problem

$$F_i(y_1 \cdots y_I) \equiv F_i(y_j) = 0, \qquad i = 1 \cdots I. \tag{18}$$

Let $\{y_j^n\}$ be the n-th approximation to the set $\{y_j\}$, and set

$$y_j^{n+1} = y_j^n + \delta_j^n .$$

Expansion of (18) in a Taylor series gives

$$F_i(y_j^n) + \sum_j \delta_j^n \frac{\partial}{\partial y_j} F_i(y_j^n) + \cdots = 0 . \tag{19}$$

Eq. (19) can be solved through first order in δ_j^n by a linear equation solver, and the process iterated until convergence.

Henyey and Wilets[9] utilized the Newton-Raphson principle to develop a method for solving coupled nonlinear differential equations subject to boundary conditions, and applied it first to the problem of stellar structure. Let the system of equations be

$$y_i'(x) + f_i\big(y_j(x)\big) = 0 . \tag{20}$$

In analogy with the algebraic case, let $\{y_i^n(x)\}$ be the n-th approximation to the set $\{y_i(x)\}$, and set

$$y_i^{n+1}(x) = y_i^n(x) + \delta_i^n(x) . \tag{21}$$

Expanding the dependent variables about each point x, we have

$$\delta_i^{n\prime}(x) + \sum_j \delta_j^n \frac{\partial}{\partial y_j} f_i(y_j^n(x)) = - \left[y_i^{n\prime}(x) + f_i(y_j^n(x))\right] + \cdots . \tag{22}$$

Ignoring higher order terms, (22) is a set of coupled, first order, linear, inhomogeneous differential equations, subject to appropriate boundary conditions. They can be solved for the $\delta_i^n(x)$, the y_i^{n+1} obtained, and the process iterated until convergence. Convergence is quadratic—the number of significant figures doubles with each iteration.

The essential feature of the method is that the boundary conditions are satisfied at every iteration. The equations are very stable so long as one is in the neighborhood of a solution. The principle here is what I call "the girl with a curl" syndrome:

There was a little girl,
And she had a little curl
Right in the middle of her forehead.
When she was good,
She was very, very good,
And when she was bad she was horrid.

H. W. Longfellow

The set (22) can be solved by standard differential equation solver routines. One can (for example) obtain one particular solution of the inhomogeneous differential equation which satisfies the inside boundary conditions and I solutions of the homogeneous differential equations which also satisfies the inside boundary conditions. The most general solution is a linear combination of the particular solution and an arbitrary linear combination of the homogeneous solutions. This gives I constants. The same can be done satisfying the outside boundary conditions. Now we have $2I$ constants. Fitting the inside and outside functions and their derivatives at an intermediate match point fixes the $2I$ constants. The programs COLSYS[10] and LARRY (Carson[11]) utilize this principle but differ in the numerical techniques.

An alternative method is to express the differential equations as difference equations on a fixed grid. Solution of the inhomogeneous difference equations involves the inversion of a band matrix, for which well-known techniques are available [cf. refs. (4) and (8)].

Different procedures have been used for solving the full set of equations (17). Goldflam and Wilets[4,8] solve the eigenvalue equations (17 a, b), assert the normalization condition (17 d), and then solve the nonlinear sigma equation (17 c). The overall process is iterated, with a convergence (or damping) factor of 0.5 on the sigma increment, to the point of convergence.

Alternatively, Saly[12], Saly and Sundaresan[13] , Köppel and Harvey[14] have employed variations of the following. Consider (17 a, b, c) to be a set of four nonlinear equations to be solved simultaneously. For fixed g_0 and ϵ, a solution can be obtained which, however, does not satisfy the normalization condition (17 d). Then a search can be made on ϵ until (17 d) is satisfied. Or, by scaling, one can interpret any solution of (17 a, b, c) as a normalized solution with *some* coupling constant g_0. This is particularly useful if one is scanning a parameter set.

Dodd and Lohe[15] cleverly are able to fully specify the model parameters and solve the full set of equations simultaneously. They do this by augmenting the variables with a normalization function, say

$$\xi(r) = \int_0^r [(u^2(r') + v^2(r')]r'^2\,dr' \,. \tag{23 a}$$

Then

$$\xi' = r^2(u^2 + v^2)\,, \tag{23 b}$$

with $\xi(0) = 0$ and $\xi(\infty) = 1$. The eigenvalue is also introduced as a variable, $\epsilon = \epsilon(r)$, satisfying the deceptively trivial differential equation

$$\epsilon'(r) = 0 \tag{24}$$

which is nevertheless a member of the coupled set. Then these 6 coupled, nonlinear, first order equations are solved simultaneously. From my own experience, I would judge that this probably requires the order of ten to twenty iterations, depending on the quality of the initial guess.

NONSPHERICAL MEAN FIELD SOLUTIONS

The first obvious extension of these system of equations is to nonspherical systems, such as are encountered in hadron-hadron scattering. One way of handling such systems is by expansion in spherical harmonics. Consider the case of axial and reflectional symmetry. The σ-field can be expanded in a set of even Legendre polynomials,

$$\sigma(\vec{r}) = \sum_l \sigma_l(r) P_l(\cos\theta)\,. \tag{25}$$

This multiplies the number of the sigma equations (17 c) by $\frac{1}{2}l_m + 1$, where l_m is the maximum order of l. The Dirac function becomes a linear combination of κ-eigenfunctions:

$$\psi_n = \sum_{\kappa} \begin{pmatrix} u_n^{\kappa}(r) \\ i\vec{\sigma}\cdot\hat{r}\; v_n^{\kappa}(r) \end{pmatrix} \mathcal{Y}_{\kappa m} \,. \tag{26}$$

Again going up to l_m in orbital states, there are $2l_m + 1$ states for $|m| = \frac{1}{2}$, decreasing with $|m|$, approximately half of which are even parity and half are odd parity. Although the different parity states do not *mix*, they do *couple*. For n spatially distinct quark states, the number of coupled, first order differential equations to be solved is the order of $l_m + 2l_m n + 2n$. Although n is usually small, like 2 or 4, l_m is characteristically 10, so we could be dealing with about 50 equations.

The mean field solutions provide a basis for generator coordinate calculations.[16] By employing a coherent state representation of the sigma field, one can derive a Schrödinger-type equation for the collective (*i.e.* interparticle) coordinate.

GLUON PROPAGATOR

In the Abelian approximation, the gluon propagator is identical to the Maxwell Green's function except for the appearance of the $SU(3)$ color matrices. In the soliton model, however, one must solve for the Green's function in a dielectric medium[17,18] which contains a spatially dependent $\kappa(\vec{r})$. This again involves the solution of coupled differential equations. The one gluon exchange magnetic interaction can be treated in perturbation theory, so I will not consider it further here.

Of special interest is the electric interaction, since this is the mechanism of color confinement. Fai, Perry and I[5] have considered the case where the model quark-sigma coupling, Eq.(5), has been set equal to zero. For vanishing current quark mass, the resulting Lagrangian is chirally invariant. Quarks are still coupled to a sigma field indirectly through the gluon field. Because $\kappa \to 0$ in the vacuum, one can easily show that quarks within a cavity, where $\kappa \neq 0$, have infinite field energy unless they are in a color singlet state. The field energy due to a single quark is its self energy.

Calculation of the quark self energy gives rise to two kinds of divergences. One is the infrared divergence, which gives absolute color confinement as just mentioned. When quarks are in a color singlet state, the infrared self energy divergence is cancelled by an infinite term in the mutual interaction. For numerical purposes, it is convenient to regulate the infrared divergence by setting the vacuum value of the dielectric function equal to a small number which is then taken to be arbitrarily small: $\kappa(\sigma_v) = \epsilon \to 0$.

The ultraviolet divergence occurs even for the case of a homogeneous medium, *i.e.*, $\kappa = 1$. For classical particles, it diverges as the inverse of a size parameter, say a, and relativistically it diverges as $\ln a$. Because of the running of the strong coupling constant, in QCD it diverges only as $\ln(\ln a)$, which is very slow indeed. The divergent self energy is normally incorporated into mass renormalization. That is is not possible here because the the divergent term is spatially dependent through $\kappa(\vec{r})$. For spherical systems in color singlet states, this gives rise (for small a) to an *effective* quark-sigma coupling which is proportional to $1/\kappa$. For non- spherical systems, such as colliding or fissioning bags. where one needs to make the color confinement explicit, it is necessary to calculate the quark self energy self consistently with the quark and sigma functions.

We have considered fixed (*i.e.* infinitely massive) quarks, although there is reason to believe that this is not so bad for light, confined, quarks. In order to regularize the short distance (ultraviolet) behavior, a form factor for the quark charge is introduced: $f(|\vec{r} - \vec{r}'|)$. It is normalized to $\int d^3r f(r) = 1$. For example, if we choose

$$f(r) = \frac{e^{-r^2/a^2}}{a^3\pi^{3/2}}, \tag{27}$$

then the divergent term can be shown to be

$$\tfrac{2}{3}\alpha_s \frac{1}{\kappa} \int d^3r f(r)/r = \frac{4\alpha_s}{3\sqrt{\pi}a\,\kappa}. \tag{28}$$

The self energy is given by

$$E_{self}(r) = \tfrac{2}{3} g_s^2 \int d^3r' G(\vec{r}, \vec{r}\,') \frac{f(|\vec{r} - \vec{r}'|)}{a^3\pi^{3/2}}, \tag{29}$$

In transverse gauge $\vec{D} = \kappa\vec{E}$ satisfies Gauss's law,

$$\vec{\nabla} \cdot \vec{D} = \rho. \tag{30}$$

The scalar Green's function $G_{00} = G^{00} \equiv G$ satisfies the time-independent differential equation

$$-\vec{\nabla} \cdot \kappa(\vec{r})\vec{\nabla}G(\vec{r},\vec{r}') = \delta^3(\vec{r}-\vec{r}'). \tag{31}$$

To simplify this equation one can define

$$\overline{G}(\vec{r},\vec{r}') \equiv \sqrt{\kappa(\vec{r}\,)}\,G(\vec{r},\vec{r}')\sqrt{\kappa(\vec{r}')}, \tag{32}$$

where $\overline{G}$ satisfies

$$(-\nabla^2 + W(\vec{r}))\overline{G}(\vec{r},\vec{r}') = \delta^3(\vec{r}-\vec{r}'), \tag{33}$$

with the "potential" $W(\vec{r})$ given by

$$W(\vec{r}) \equiv \tfrac{1}{4}|\vec{\nabla}\ln\kappa(\vec{r})|^2 + \tfrac{1}{2}\nabla^2\ln\kappa(\vec{r}). \tag{34}$$

We expand $W(\vec{r})$ and $\delta^3(\vec{r}-\vec{r}')$ into spherical harmonics,

$$W(\vec{r}) = W_{LM}(r)Y_{LM}(\Omega), \tag{35}$$

$$\delta^3(\vec{r}-\vec{r}') = \frac{1}{r^2}\delta(r-r')Y_{LM}(\Omega)Y^*_{LM}(\Omega'), \tag{36}$$

where the repeated index summation convention is used.

For the scalar Green function we make the ansatz

$$\overline{G}(\vec{r},\vec{r}') = C_{\alpha\alpha'}\frac{1}{rr'}f^{\alpha}_{lm}(r_<)Y_{lm}(\Omega_<)g^{\alpha'}_{l'm'}(r_>)Y^*_{l'm'}(\Omega_>), \tag{37}$$

where the quantities $(r_<,\Omega_<,i_<)$ refer to (r,Ω,i) if $r > r'$ and to (r',Ω',i') if $r < r'$; $(r_>,\Omega_>,i_>)$ is defined correspondingly. The radial functions $f^{\alpha'}_{lm}(r)$ and $g^{\alpha}_{lm}(r)$ satisfy the coupled differential equations

$$\left\{\left(-\frac{d^2}{dr^2} + \frac{l(l+1)}{r^2}\right)\delta_{ll'}\delta_{mm'} + W_{LM}(r) < lm \mid Y_{LM} \mid l'm' >\right\}$$

$$\times\left\{\begin{array}{c} f^{\alpha}_{l'm'}(r) \\ g^{\alpha'}_{l'm'}(r)\end{array}\right\} = 0. \tag{38}$$

The set $\{\alpha\}$ of solutions, which are regular at the origin, is given by $\{f^{\alpha}_{lm}\}$ and the set $\{\alpha'\}$, which are regular at infinity, by $\{g^{\alpha'}_{l'm'}\}$. If

$\kappa(\vec{r})$ goes asymptotically to a *constant greater than zero*, the corresponding boundary conditions are

$$f^{\alpha}_{lm}(r) \sim r^{l+1}\delta^{\alpha}_{lm} \qquad \text{for } r \to 0, \tag{39 a}$$

$$g^{\alpha'}_{l'm'}(r) \sim \frac{1}{r^{l'}}\delta^{\alpha'}_{l'm'} \qquad \text{for } r \to \infty. \tag{39 b}$$

If there are no symmetries in the problem, all (l, m) are coupled. In an axially symmetric problem, the potential W can be expanded in terms with the magnetic quantum number $M = 0$ only, and the differential equations (38) decouple for different m. For reflectional symmetry, only even L appear. The coefficients $C_{\alpha\alpha'}$ in (37) satisfy the linear algebraic equations

$$\left\{f^{\alpha}_{lm}(r)n^{\alpha'}_{l'm'}(r) - j^{\alpha}_{l'm'}(r)n^{\alpha'}_{lm}(r)\right\}C_{\alpha\alpha'} = 0, \tag{40 a}$$

$$\left\{\left(\frac{d}{dr}f^{\alpha}_{lm}(r)\right)n^{\alpha'}_{l'm'}(r) - j^{\alpha}_{l'm'}(r)\left(\frac{d}{dr}n^{\alpha'}_{lm}(r)\right)\right\}C_{\alpha\alpha'} = \delta_{l,l'}\delta_{m,m'}, \tag{40 b}$$

at any radius r. Equations (40 b) with $l < l'$ are redundant, so that (40) reduces to the linearly independent set of equations: $l < l'$ for (40 a) and $l \geq l'$ for (40 b). After solving the linear equations for $C_{\alpha\alpha'}$, the scalar Green function is completely determined.

The effective selfconsistent nucleon (3 quarks) mean field equations can be summarized by the following coupled nonlinear integro-differential equations:

$$\left[\vec{\alpha}\cdot\vec{p} + \beta\left(m + E_{self}(\vec{r}) - \frac{4\alpha_s}{3\sqrt{\pi}a}\right) + V_n(\vec{r}) - \epsilon_n\right]\psi_n(\vec{r}) = 0\,, \tag{41}$$

$$-\nabla^2\sigma + U'(\sigma) + 2g_s^2\kappa'(\sigma)\sum_{n(valence)}\left[\int \vec{\nabla}G(\vec{r},\vec{r}\,')\,|\psi_n(\vec{r}\,')|^2 d^3r\,'\right]^2 = 0\,, \tag{42}$$

where E_{self} is defined in Eq. (29) and $G(\vec{r},\vec{r}\,')$ by (37). The term $4\alpha_s/(3\sqrt{\pi}a)$ is the quark mass renormalization. The exchange potential V_n, as in Hartree-Fock, can be written as a nonlocal operator,

$$V_n(\vec{r},\vec{r}\,') = \sum_{n'(\neq n)} <n|\vec{\lambda}|n'> \cdot <n'|\vec{\lambda}'|n> \psi_n(\vec{r})G(\vec{r},\vec{r}\,')\psi^{\dagger}_{n'}(\vec{r}\,')\,. \tag{43}$$

These equations are coupled, nonlinear, integro-differential equations. When placed on a grid, the Newton-Raphson linearization leads requires the solution of (or inversion of) a matrix equation where the matrix is no longer a band matrix, but involves a relatively full matrix of large dimension, governed by the spatial grid. This is clearly computer intensive.

OTHER TOPICS

There are many other aspects of QCD modelling with solitons which have not been detailed above. Among the computer-intensive ones are:

The solution of the dynamical soliton equations utilizing the generator coordinate method with coherent states.

The many bag problem, which has been attacked by the Wigner-Seitz approximation.[19]

Vacuum polarization, or loop corrections.[20] These are the further quantum corrections to the mean field approximation, and involve calculating the negative energy Dirac sea for the quarks and zero-point oscillations for the sigma field.

REFERENCES

1. R. Friedberg and T. D. Lee, Phys. Rev. **D15**, 1694 (1977); **D16**, 1096 (1977); **D18**, 2623 (1978).
2. T. D. Lee, *Particle Physics and Introduction Field Theory* (Harwood Academic, New York, 1981).
3. K. Huang and D. R. Stump, Phys. Rev. **D 14**, 223 (1976).
4. R. Goldflam and L. Wilets, Phys. Rev. **D25**, 1951 (1982); L. Wilets, *Nontopological Solitons* (World Scientific, Singapore, to be published).
5. G. Fai, R. Perry and L. Wilets, "The Chromo-dielectric Model: Confinement, effective coupling and chiral invariance," Lawrence Berkeley Laboratory preprint, LBL-24046 (1988).
6. M. Bickeböller, 'Der Gluon-Propagator im Soliton-Bag-Modell,' Diplom Thesis, University of Bonn, 1984;
   M. Bickeböller, R. Goldflam and L. Wilets, J. Math. Phys. **26**, 1810 (1985).

7. A. Schuh, 'Nukleon-Nukleon-Wechselwirkung im Soliton Bag Modell,' Doctoral Dissertation, University of Heidelberg (1985); "Dynamics of 6 Quark Systems in the Soliton Bag Model;" A. Schuh, H. J. Pirner and L. Wilets, Physics Letters B. **174**, 10 (1986).
8. R. Horn, R. Goldflam and L. Wilets, Comp. Phys. Com. **42**, 105 (1986) .
9. L. Wilets and R. A. Berg, Phys. Rev. **101**, 201 (1956); L. G. Henyey, L. Wilets, K. H. Böhm, R. Le Levier and R. D. Levee, Astrophys. J. **129** (1959).
10. U. Ascher, J. Christiansen and R. D. Russell, Math. Comp. **33**, 659 (1979); *Lecture Notes in Computer Science*, **76**, (Springer, N. Y.); (1981) ACM. Trans. Math. Software **7**, 223 (1981).
11. L. C. Carson (private communication).
12. R. Saly, Comput. Phys. Commun. **30**, 411 (1983).
13. R. Saly and M. K. Sundaresan, Phys. Rev. **D29**, 525 (1984).
14. Th. Köppel and M. Harvey, Phys. Rev. **D 31**, 171 (1985).
15. L. R. Dodd and M. A. Lohe, Phys. Rev. **D 32**, 1816 (1985).
16. L. Wilets, "The Non-topological Soliton Bag Model," in *Chiral Solitons*, edited by K.-F. Liu (World, Singapore), 362 (1987).
17. M. Bickeböller, 'Der Gluon-Propagator im Soliton-Bag-Modell,' Diplom Thesis, University of Bonn (1984); M. Bickeböller, R. Goldflam, R. and L. Wilets, J. Math. Phys. **26**, 1810 (1985).
18. M. Bickeböller, M. C. Birse, H. Marschall and L. Wilets, Phys. Rev. **D 31**, 2892 (1985).
19. M. Birse, J. J. Rehr and L. Wilets, submitted for publication (1988).
20. Ming Li, Doctoral Dissertation, University of Washington (1987); Ming Li, R. J. Perry and Lawrence Wilets, Journal of Computational Physics to be published.

CONDENSATES IN THE STANDARD MODEL

CARL SHAKIN
Department of Physics, Brooklyn College of CUNY, Brooklyn, N.Y. 11210

Abstract: We discuss the role of the gluon condensate in chiral symmetry breaking and in the confinement of quarks. We describe a covariant mean-field theory of confined quarks (covariant soliton dymanics). We also comment on the possible observation of a confined phase of electroweak theory.

INTRODUCTION

In this contribution we stress the importance of the gluon condensate in understanding the dynamics of confinement and mass generation for light quarks (chiral symmetry breaking)[1-2]. We find that the dynamical mass is strongly momentum dependent; the construction of a covariant theory of hadron structure, in which the hadrons consist of confined quarks with strongly momentum-dependent dynamical masses, is seen to be a challenging problem requiring rather complex calculations.

We also comment upon the recent observation of sharp peaks in the e^+-e^- spectra observed in large Z heavy-ion collisions[3] and speculate that these electrons and positrons may have their origin in the decay of nontopological solitons in a confined phase of electroweak theory.[4-5]

In the limited space available for this discussion we are not able to present any details of our work, nor are we able to provide complete references to the extensive literature concerning the matters under discussion. References to the work of others and further details may be found in the references given here.

GLUON CONDENSATES IN QCD

It is generally thought that the QCD ground state contains a gluon condensate[6]. Any discussion of this issue raises the question of the gauge invariance of the analysis. For our analysis we work in some (preferably covariant) fixed gauge. We have made the assumption that the gluon condensate is in a zero-momentum mode and have constructed a model in which the local gauge symmetry of QCD is spontaneously broken down to a global symmetry when the condensate is formed. We have shown, within a specific mean-field analysis, that quarks (of any current mass) do not propogate in the gluon condensate.[1-2] We also find dynamical mass generation (chiral symmetry breaking) for massless quarks in a nonperturbative analysis. In addition, we find that the gluon condensate leads to dynamical mass generation for gluons *via* the Schwinger mechanism. Further, we see that gluons do not propagate in the gluon condensate.[7-8] We will summarize some of the essential features of our results before considering the problem of describing hadron structure in terms of condensate order parameters.

We may start with our result for the gluon propagator[7-8] written in terms of an order parameter of the gluon condensate, φ_o^2,

$$iD_{\mu\nu}^{ab}(k) = -\varphi_0^2 \frac{g^{\mu\nu}}{4} \frac{\delta_{ab}}{8} (2\pi)^4 \delta^4(k)$$
$$-i \left[\frac{g_{\mu\nu}-k_\mu k_\nu/k^2}{k^2\left[1-\Pi_1(k^2)\right]} + \frac{\lambda k_\mu k_\nu}{(k^2+i\epsilon)^2}\right]\delta^{ab} . \tag{1}$$

Here a,b are color indices and λ is a gauge-fixing parameter. The quantity $\Pi_1(k^2)$ is related to the vacuum polarization tensor,

$$\Pi_{\mu\nu}^{ab}(k) = \delta^{ab}(g_{\mu\nu}-k_\mu k_\nu/k^2)\Pi(k^2) \quad , \tag{2}$$

$$= \delta^{ab}(g_{\mu\nu}k^2-k_\mu k_\nu)\Pi_1(k^2) \quad . \tag{3}$$

The first term in Eq. (1) describes the zero-momentum condensate of gluons. In our earlier work[9], we related $g^2\varphi_o^2$ to the vacuum expectation value

$$<vac|\frac{g^2(\tilde{\mu}^2)}{4\pi^2} G_{\mu\nu}^a(0)G_a^{\mu\nu}(0)|vac> = 0.012(\text{GeV})^4 \quad , \tag{4}$$

whose value is given in QCD sum-rule studies.[6] This quantity vanishes in the perturbative vacuum and we ascribe the nonzero value to the presence of a zero-momentum condensate.

The dielectric constant of the QCD vacuum may be defined as[8]

$$\kappa(k^2) = 1 - \Pi_1(k^2) \ . \tag{5}$$

If $\Pi_1(k^2)$ has pole at $k^2=0$, the transverse gluons obtain a dynamical mass. This is known at the "Schwinger mechanism"[10] and is analogous to the fact that the photon becomes massive in a superconductor leading to the Meissner effect. Further, if $\kappa(k^2)$ has no zeroes, the gluon propagator has no pole. The gluon is therefore a <u>nonpropagating mode</u>. (We prefer this terminology to saying that the gluon is "confined".) Our result for $\Pi^{ab}_{\mu\nu}(k)$ is [7-8]

$$\Pi^{ab}_{\mu\nu}(k) = \delta^{ab}\left(g_{\mu\nu} - k_\mu k_\nu/k^2\right)\left[m_G^2 - \frac{4\eta^2 k^2}{k^2-m_G^2}\right] , \tag{6}$$

where $m_G^2=3\eta^2=(9/32)g^2\varphi_o^2$. The first term corresponds to the action of the Schwinger mechanism, while the second term gives rise to confinement, as can be seen in Fig. 1 where we present $\kappa(k^2)$ as a function of k^2.

Note that we have not included the perturbative QCD corrections to this result; we have $\kappa(k^2) \to 1$ for $k^2 \to \pm\infty$. We see from the figure how the second pole in $\kappa(k^2)$ at $k^2=m_G^2$ leads to the fact that $\kappa(k^2)$ has no zeroes. Thus the gluon is <u>massive</u> and <u>nonpropagating</u>.

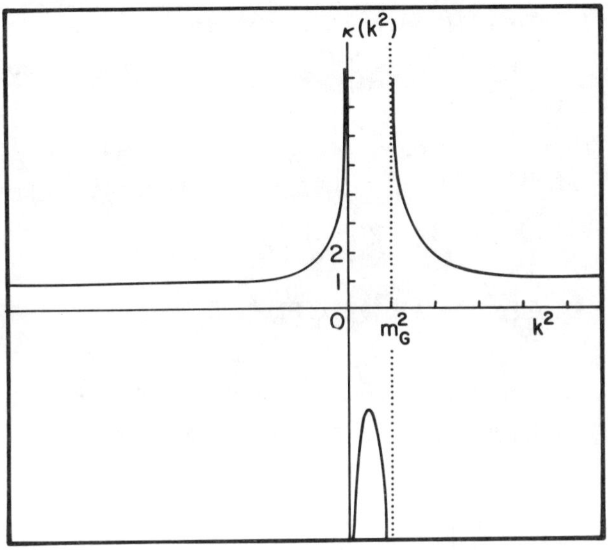

FIGURE 1. Dielectric constant of the QCD vacuum (nonperturbative aspects).

Now let us ask for the behavior of the quark propagator in the gluon condensate. We concentrate on the semi-classical aspects of this problem and write a Schwinger-Dyson equation for the quark self-energy using only the first term of Eq. (1), which describes the zero-momentum gluon condensate. (Again, we are working in some fixed covariant gauge and we note that the quark propagator is a gauge-dependent quantity.) The quark propagator is

$$S_F(p) = \left[\not{p} - m_q^{cur} - \Sigma(p)\right]^{-1} \tag{7}$$

where $\Sigma(p)$ is the self-energy due to the presence of the condensate. Here m_q^{cur} is a "current" quark mass, whose origin is to be found in a theory which is more general than QCD. (For up and down quark, $m_q^{cur} \sim 5$ MeV; for the charm quark, $m_q^{cur} \sim 1.4$ GeV, while for the bottom quark $m_q^{cur} \sim 5$ GeV.) We write

$$\Sigma(p^2) = A(p^2)\not{p} + B(p^2) \tag{8}$$

and present an equation for $\Sigma(p)$ in Fig. 2. There the cross-hatched region denotes the condensate and the

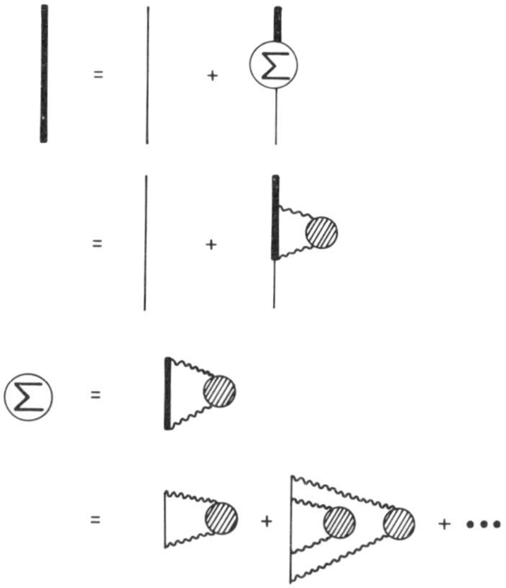

Figure 2. Self-consistent equations for the quark self-energy and quark propagator.

wavy lines are zero-momentum gluons. The heavy line is the quark propagator of Eq. (7). One has

$$A(p^2)\not{p} + B(p^2) = \frac{-\kappa^2\gamma^\mu\left[\not{p}(1-A(p^2)) + B(p^2)\right]\gamma_\mu}{p^2(1-A(p^2))^2 - B^2(p^2)} \qquad (9)$$

where $\kappa^2 = g^2\varphi_o^2/24$. If $m_q^{cur} = 0$ one can obtain an analytic solution

$$\left.\begin{aligned} A(p^2) &= -1 \\ B(p^2) &= 2\sqrt{\kappa^2 + p^2} \end{aligned}\right\} \quad p^2 > -\kappa^2$$

$$\left.\begin{aligned} A(p^2) &= \frac{1}{2}\left[1 - \sqrt{1-8\,\kappa^2/p^2}\,\right] \\ B(p^2) &= 0 \end{aligned}\right\} \quad p^2 < -\kappa^2 \qquad (10)$$

which exhibits chiral symmetry breaking since $B(p^2)\neq 0$ (even with $m_q^{cur}=0$). It is useful to define a dynamical mass[2]

$$M(p^2) = \frac{B(p^2) + m_q^{cur}}{1 - A(p^2)} \quad . \qquad (11)$$

In Fig. 3 we show $M(p^2)$ for a light quark and in Fig. 4 we show $M(p^2)$ for a heavy quark. In both cases, for large timelike p^2, we have $M^2(p^2) \rightarrow p^2+\kappa^2$. Note that as long as $\kappa^2\neq 0$, the quark is a nonpropagating mode in the gluon condensate.

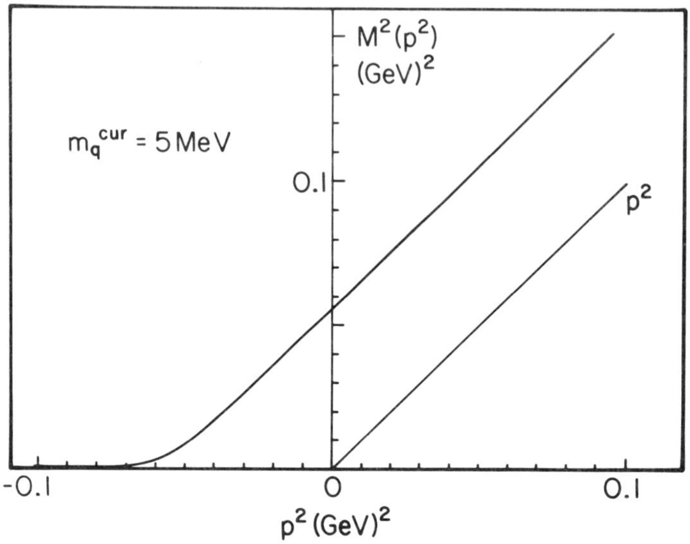

Figure 3. Dynamical mass of a light quark

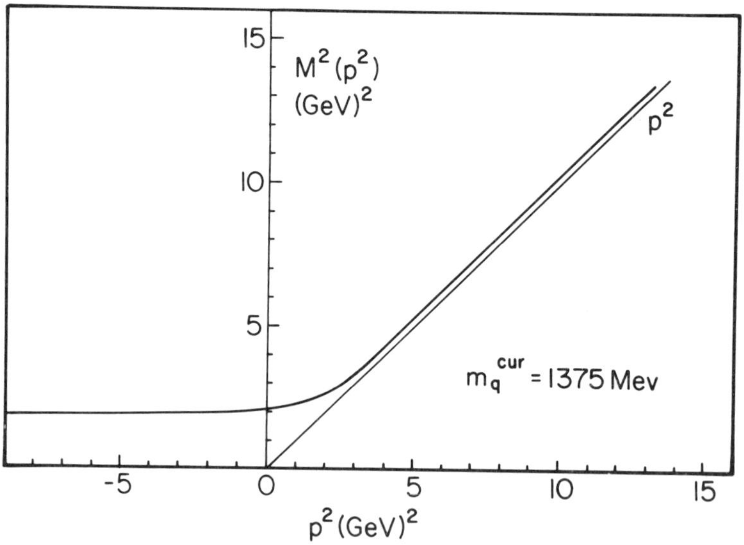

Figure 4. Dynamical mass of a heavy quark

COVARIANT SOLITON DYNAMICS

Let us now turn to our theory of hadron structure (covariant soliton dynamics). The basic idea underlying this and similar models is that the presence of quarks destroys the condensate structure locally and the quarks

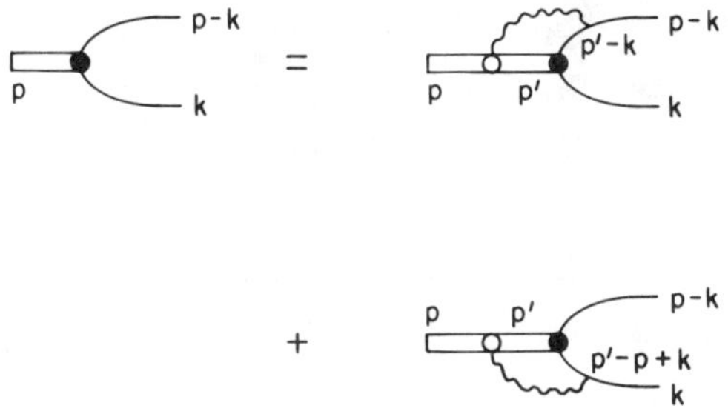

Figure 5. Mean-field equation governing meson structure.

move in a region where the condensate order parameter has a smaller value. A covariant mean-field analysis leads to nonlinear equations for quark vertex functions[11-12].

In Figure 5 the filled circle is a vertex function and the open circle denotes a form factor which is a functional of the vertex function. We cannot discuss equations of the theory in any detail here; however, we note that the vertex is a function of two variables, for example, p•k and k^2. For a pseudoscalar meson, the vertex function has the form

$$\Gamma(p\bullet k,k^2) = \gamma_5\Big[a_1(p\bullet k,k^2)+a_2(p\bullet k,k^2)\not{p} + a_3(p\bullet k,k^2)\not{K} + a_4(p\bullet k,k^2)\not{p}\not{K}\Big] . \qquad (12)$$

Thus, in this case, there are <u>four</u> unknown functions of <u>two</u> variables. These functions satisfy nonlinear equations[13] which may be solved by iteration. This is clearly a problem which is quite demanding of computer resources. Simplifications, such as placing one quark on mass shell[11], cannot be made, since, if we include quark self-energies of the type discussed above, the quarks cannot go on mass shell[12-13]. (It is worth noting that, because of the form of the quark self-energy, our equations have only bound-state solutions.)

ELECTROWEAK CONDENSATES

There is a rather remarkable body of data emerging from the study of large Z heavy-ion collisions.[3] The observation of sharp peaks in the electron and positron spectrum has been followed by the observation of quite narrow peaks in the photon spectrum[14] arising in heavy-ion collisions. It has been suggested that one is seeing the decay of a quite large object (a nontopological soliton), possibly 300-3000 fm in radius, formed in a new strongly coupled phase of QED.[4-5, 15] Indeed, it is possible to show, using the same techniques as described above, that if a photon condensate is formed, electrons and positrons will be nonpropagating modes.[16] We can then use the techniques of covariant soliton dynamics to describe the formation of nontopological solitons in the confined phase.

One may object that the coupling constant in QED is still small even in the presence of large fields and, therefore, a discussion of strongly coupled QED may be irrelevant.[17] One is therefore led to develop a scenario[16] in which the effective theory, which is appropriate to describe the physics a new phase, is similar to strongly-coupled QED. One way to discuss this question is within the context of electroweak theory with two Higgs doublets. If the second Higgs doublet were to become massless under certain circumstances, the interaction of the electromagnetic fields and the Higgs fields would have strong infrared singularities. If one were to "integrate out" the Higgs field, one might be left with a theory which has some aspects of strongly-coupled QED. These ideas are, of course, quite speculative, however, the experimental data is such as to lead to such speculations. Further work is required, both in theory and experiment.

REFERENCES

1. V.M. Bannur, L.S. Celenza and C.M. Shakin, Brooklyn College Report: BCCNT 87/101/170 (1987) - submitted for publication.

2. V.M. Bannur, L.S. Celenza and C.M. Shakin and Hui-Wen Wang, Brooklyn College Report: BCCNT 87/102/171 (1987) - submitted for publication.
3. For a review see, T.E. Cowan and J.S. Greenberg et al., in Physics of Strong Fields, edited by W. Greiner (Plenum Press, New York, 1987).
4. L.S. Celenza, V.K. Mishra, C.M. Shakin and K.F. Liu, Phys. Rev. Lett. 57, 55 (1986).
5. L.S. Celenza, Chueng-Ryong Ji and C.M. Shakin, Phys. Rev. D36, 2144 (1987).
6. See, for example, M.A. Shifman, Annu. Rev. Nucl.Part. Sci. 33, 199 (1983).
7. L.S. Celenza, C.M. Shakin and Hui-Wen Wang, Brooklyn College Report: BCCNT 87/111/172 (1987) - submitted for publication.
8. L.S. Celenza, C.M. Shakin and Hui-Wen Wang, Brooklyn College Report: BCCNT 87/121/173 (1987) - submitted for publication.
9. L.S. Celenza and C.M. Shakin, Phys. Rev. D34, 1591 (1986).
10. J. Schwinger, Phys. Rev. 125, 397 (1962).
11. L.S. Celenza, A. Rosenthal and C.M. Shakin, Phys. Rev. C 32, 212 (1985); 32, 232 (1985).
L.S. Celenza, C.M. Shakin and R.B. Thayyullathil, Phys. Rev. D 33, 198 (1986).
12. V.M. Bannur, L.S. Celenza and C.M. Shakin, Brooklyn College Report: BCCNT 87/071/168 (1987) - submitted for publication
13. V.M. Bannur, L.S. Celenza and C.M. Shakin, Brooklyn College Report: BCCNT 87/091/169 (1987) - submitted for publication.
14. K. Danzmann et al., Phys. Rev. Lett. 59, 1885 (1987).
15. D.G. Caldi and A. Chodos, Phys. Rev. D 36, 2876 (1987).
Y.J. Ng and Y. Kikuchi, Phys. Rev. D 36, 2880 (1987).
See also, W. Greiner in Physics of Strong Fields, edited by W. Greiner (Plenum Press, New York, 1987).
16. L.S. Celenza, A. Pantziris, C.M. Shakin and Hui-Wen Wang, Brooklyn College Report BCCNT: 88/031/175 (1988) - submitted for publication.

17. R.D. Peccei, J. Solà and C. Wetterich, DESY pre-
print (Dec. 1987).

A Tutorial on Distributed Memory Programs

D. E. Stevenson,

Department of Computer Science,
Clemson University,
Clemson, SC 29634-1906

ABSTRACT

Computers such as the Floating Point Systems T-series are a significant departure from Vax and CRAY computers. There are a large number of concepts which must be understood in order to read the literature. Furthermore, the ideas behind program development are different from the current structured program methodology. We present an overview of both the hardware and software concepts in distributed computing.

Introduction

At the suggestion of the session chair, this tutorial attempts to pull together various aspects of distributed programming concepts. The goal is to provide the reader with a overview. The references chosen are not necessarily the latest. Rather, the author has tried to choose them based on their elementary nature and ready availability.

The approach is to start with the hardware and work up to distributed memory systems and their programming. There are six sections:

1. Basic hardware concepts.
2. Basic concepts in programming languages.
3. Foundations of parallel programming.
4. Concepts in synchronization.
5. Message passing for distributed systems.
6. Distributed programs.

In the short space suitable for a proceedings, the author has tried to give the reader a very broad basis; hopefully, the reader will excuse the lack of depth.

1. Some Background Information on Computer Architecture Terminology

Since we must make some fine distinctions about computer architecture, it is worth discussing a high level abstraction. Such abstractions occur in the literature, but we will use two ready references: Siewiorek, Bell, and Newell[11] along with Reddi and Feustel[10] . From Reddi and Feustel we present one possible abstraction for architecture. There are three basic considerations:

M1 The physical organization of the computer,

M2 The control and flow of information, and

M3 The representation, interpretation, and transformation of information.

We briefly look at each, eventually to distinguish hypercubes and more familiar computers. These above themes recur throughout the following sections.

Physical organization. We may abstract the physical organization into seven components[11] . These are

M1a *Memory* and the attendant *addressing schemes.*

M1b *Control* which evokes operations in other components.

M1c *Transducers* which alter data forms, but which will not be of interest to us here.

M1d *Data operations* which have the obvious use.

M1e *Processors* which are capable of interpreting programs.

M1f *Links* which transfer information among components.

M1g *Switches* which allow the setting of links.

A reasonable way to abstract the organization itself is to consider a computer as a graph with vertices from the set of components and edges as the flow of information from one component to another.

Control and Flow of Information. This abstraction deals with the manner of processing. Such technical issues as the sequencing of control and data operations, interrupts, and memory addressing fall in this category. In the current literature, one often sees the Flynn[5] taxonomy of control of processing streams:

- *SISD:* A *single instruction* processes a *single operand set.* Examples are the VAX and PC's.
- *SIMD:* A *single instruction* processes a *multiple operand sets.* Examples are the CRAY-1 and Connection Machine.
- *MISD:* A *multiple instructions* processes a *single operand set.* There are no examples of this class.
- *MIMD:* A *multiple instructions* processes a *multiple operand sets.* Examples are multiprocessors and many special architectures like hypercubes. See **P1** below.

Representation, Interpretation and Transformation of Information. While the first two concepts support this third concept, we as programmers are most interested in the primitive data operations available to us. We do not belabor the point, other than to note that the question is what scalar and aggregate data types are directly supported with what operations and relations.

The plethora of new machines is being forced by solutions to the so-called *von Neumann bottleneck.* In the conventional von Neumann model, we have one *processor* and one *memory* linked by a *bus.* Since all data and control is constrained to move over the bus it is possible to have queuing delays. Each new method for eliminating these queuing delays gives rise to a new machine architecture. For example, a *shared memory multiprocessor* employs many processors each with a bus, but with only one memory. The obvious example for most readers is the CRAY Y-MP. Another solution is the *distributed memory multiprocessor* which has many processors, each with a private memory. To share information in a distributed memory multiprocessor, one uses buses between *nodes* (processor-memory pairs). How and why these buses - or *channels* - interconnect also leads to new abstract machines. For example, the *hypercube* interconnection scheme connects adjacent nodes according to the *reflected binary Gray code* scheme. This particular scheme represents the flow of data in the fast Fourier transform algorithm. Just as we have a virtual machine [see below], we can have *virtual interconnection schemes.* Such schemes would be presented to the programmer in the same way that the virtual machine is.

With all this in hand, it is now possible to describe our particular machine, the Floating Point Systems T-series hypercube:

- Our abstract machine is a distributed memory machine. Each *node* has a conventional von Neumann processor with a vector coprocessor.
- Our interconnection scheme is the hypercube scheme.
- Our virtual machine contains all the usual functions to support Fortran with vector extensions. The virtual machine allows for *multitasking* and *interprocess communications.*

2. Some Background Information on Programming Language Terminology

In line with the above section, we include a similar discussion of concepts in programming languages. A readable introduction is Pratt.[9] In order to define a programming language and construct a compiler for it, one needs to consider some basic questions:

P1 What data types will be primitive to the language? Each *data type* is a definition of how the data values are represented and what primitive functions are to be included in the language definitions. While this is related to **M3** we certainly can include types not directly supported by the hardware.

P2 What mechanism is to be used to resolve the location of data values and procedural names? Such decisions determine the *scoping rules.* This is related to **M1**. For *MIMD* and distributed architectures, this can become quite involved.[12]

P3 What are the rules for passing control among the various primitive statement types in the language [*cf.* **M2**]? These rules are by far the most difficult to describe and constitute much of what makes one language different from another. The problem is multi-faceted and can be broken down as follows:

+ Intra-expressional rules, such a operator precedence.
+ Inter-statement rules, such as **gotos**, **if-then-elses**, and **whiles**.
+ Compound-statements, like **begin-end** blocks in Pascal. We can also take this as how statements are combined into procedures.
+ Procedural linkages and argument passing. Much more on this below and
+ Modular definitions such as appear in ADA, C++, Modula-2, or SmallTalk.

Control is usually thought of as *sequential* control; in the newer architectures, however, there other forms.

P4 What memory management techniques are to be used and how much control the programmer has? While there is no direct parallel in the architectural world, some features - virtual memory, for example - are dependent on support for effective implementations.

P5 How will the programmer communicate with the "outside" world. In the past, this was simply input/output support? However, the new architectures like the hypercube rely on more exotic functions.

The above aspects of language are normally mapped onto a virtual machine. A *virtual machine* is the set of procedures and data available to the programmer as well as concept of how a program is to be evaluated. The virtual machine is thus the total computational environment presented to the programmer. The goal of most languages is to present a model which makes semantic sense while giving the programmer a model to think about computing. For example, the *C* language presents a very simple model, leaving more complex computational functions to libraries which are outside the compiler's control.

Within the virtual machine, language design often posits an *abstract machine* which models the control structures/functions in the language. The abstract machine, for example, may have the concept of a **goto** but the virtual machine may not. For the compiler, it is the abstract machine, not the virtual machine, which is the object of design. What now makes the design of languages "fun" is the generation of many new (physical) machines whose abstract machines are radically different. To illustrate: while one would not consider a personal computer and a large mainframe equal in computing power, they are effectively the same computer at the abstract machine level. Obviously, the abstract machine must eventually be mapped onto real machine; that's the compiler's job.

3. A Short Introduction to Parallel Programming Concepts.

With the above background, we now can describe a useful model for parallel processing. In this, we follow Hoare[6] and Milner[8] .

We begin by making clear a distinction in two meanings ascribed to the term *program.* The first idea is the *source program* submitted to the compiler; the second is the object which actually is used to compute with. This second is constructed by the compiler from the first. However, each time a subroutine is called there must be a marriage between a program definition and values. This marriage is referred to as an *activation.* We will use the term *process* to mean the smallest independent unit that can be run on the computer. For example, conventional mainframes usually treat a whole load module as a process. On the other hand, OCCAM[7] treats assignment statements as processes. If P is an activation, then the significant events in its lifetime are below:

1. P is *activated* by *receiving* argument values.
2. P *computes* using only working storage whose locations are only known to P.
3. P *transmits* its results and possibly loops to 2.
4. P eventually *terminates*.

For our purposes, it is useful to think of P as being run on a machine which has a uniform execution time for primitive statement types. Since there can be many process activations in the system at any one time, we can think of an *activation space* as the set of all activations. As usual, we will define the *state* of a process as the value of all variables know to the process.

Now suppose we have two process (activations) P and Q. If P and Q are do not share any variables we will call them *independent* processes. Clearly, each process may proceed at its own pace. The pair $<P, Q>$ is itself a process. A way to visualize the difference in conventional uniprocessor and multiprocessors (with or without shared memory) is to note that the state sequences or *trace*:

- for the uniprocessor would show first one process and then the other in control.
- for the multiprocessor would show the two advancing simultaneously.

For the composite system $<P, Q>$, and looking at P, we can see four types of interactions:

T1. P computes, transmits a result and terminates.

T2. Q computes, transmits a result, P is activated by receipt of the message from Q. P proceeds as in 1. Q proceeds independently.

T3. P computes, sends Q a message, Q computes, sends P a message, and P computes again.

T4. P and Q proceed independently until they must synchronize. They exchange messages, interact for some period of time, and then proceed independently.

The traces associated with **T1** processing are what most programmers experience today. These programs run with no interactions with other programs in the system. In **T2** processing, P waits for Q to produce a result; such processing is commonly called a *pipe* or a *pipeline*. Pipes are familiar to *UNIX* programmers. **T3** programs may also be familiar to *UNIX* programmers; a new independent shell can be started and completion tested. **T4** is unlikely to be familiar to the reader as it is typical of operating system programming. It is also the major concept in programming hypercubes. We explore this in the next section.

4. Principles of Synchronization

The problems of synchronization has received much study in computer science over the course of the last twenty years. The problem naturally occurs in operating systems as well as in several application areas like data bases and discrete event simulation. There are many very readable texts in operating systems. One such is Bic and Shaw[2] ; a general introduction to concurrency/synchronization can be found in Andrews and Schneider[1] . An older, but still interesting, work concentrating on performance is by Coffman and Denning[3] . The problems of synchronizing across process boundaries is one of the keys to distributed systems. There are three distinct issues:

- Mutual Exclusion. *Mutual exclusion* is the denial of access of a resource by one process to another. In this context, *resource* is anything needed by a process: files, data, *etc*.
- Synchronization. *Synchronization* is the agreement of two processes to coordinate activities. This may be for timing reasons [hardware] or logical, data-oriented reasons.
- Deadlock Detection and Prevention. *Deadlock* is the term used to describe the situation wherein a set of process are mutually awaiting an event which cannot occur. [The terminology is quite ghoulish.] The problem is well-studied and a large number of protocols exist to prevent deadlock. The situation is dynamically detectable and preventable.

Since these three problems are so fundamental, Table 1 lists the commonly discussed mechanisms. The point, of course, is that all these mechanisms should be available to the programmer. The importance of having several models of synchronization available to the programmer cannot be overemphasized, despite the current controversy over the shared/unshared memory models.

Busy Waiting Mechanism	Mutual Exclusion
Interrupts	Semaphores
Loop Busy Wait	Critical Regions
Switches	Conditional Critical Regions
Locks	Monitors
Synchronization	**Interprocess Communications**
Semaphores	Semaphores
Critical Regions	Signals - Hardware and Software.
Conditional Critical Regions	Pipes
Monitors	Ports
Event Count/ Sequences	Synchronous Message Passing
Path Expressions	Asynchronous Message Passing
Invariant Expressions.	

Table 1. Common Synchronization Methods.

5. Message Handling.

The hypercube, due to its unshared memory, is a natural candidate for using a message passing model. In the *(one-way)(synchronous) message passing model*, the problems of mutual exclusion and synchronization are solved by having the processes communicate values through the following scheme:

- The two processes share a one-way communications channel.
- When a process wishes to transmit values, a *send* operation gains control of the channel. If the receiving process is "listening," the message is passed; otherwise, the sender must wait.
- When a process must receive values, a *receive* operation "listens" to the one-way channel. If the sending process is attempting to send, the message is passed; otherwise, the receiver waits.

Although message passing has been of theoretical interest in operating systems, it becomes the mainstay synchronization technique in distributed systems. It suffices to list these considerations:

- The implemented topology *versus* the natural N-space polyhedra of the problem. This is a real performance issue. Certainly no system would be able to have available all possible patterns. Hence, the language should take up the slack.
- Both adaptive and fixed communication algorithms can occur. Even if a pattern is fixed for a particular phase, the patterns may be radically different between phases.

- It is desirable to have localized communications wherever possible. Only some sort of development system capable of analyzing a program could optimize these patterns.

6. Issues in Designing Distributed Algorithms

Finally!

With all this mechanism in hand, we can now look at the issues of developing an algorithm for a hypercube. There are six basic issues which build on the previous language principles:

- Process creation and removal.
- Control of the progress of processes. [No "starvation."]
- Exceptional condition handling.
- Allocation and deallocation of resources.
- Interprocess communications.
- Interprocess synchronization.

The problem of interprocess relationships is perhaps new to the reader. The basic question revolves around the concept of sequentiality. A *sequential* or *deterministic* process is one whose individual behavior is not changed because of the particular set of processes it runs in [its *trace history*]. There are such things as *nondeterministic* algorithms[4] but we will not explore them here. Further considerations are

- Avoidance of deadlock.
- Data to share or deny, and by what protocol. And
- Message passing protocol.

Epilogue

We have taken a quick look at the background and issues in distributed systems. There are many changes which one needs to make in thinking about algorithms:

- Physical layout of data may span many machines.
- Communications among subroutines is more complicated.
- Control of independent processes is more complex and fraught with new dangers.
- Programming language design lags behind hardware capabilities by a wide margin.

Despite all these concerns, parallel processing may reduce the run time requirements of sequential programs by impressive amounts. While it is an open question as to what algorithms can be effectively parallelized, it is clear that not every algorithm may reap the purported benefits. As with any tool, one must evaluate the tool in light of the problem. As always, *caveat programmeris.*

References

1. Gregory R. Andrews and Fred B. Schneider, ''Concepts and Notations for Concurrent Programming,'' *Computing Surveys*, vol. 15, no. 1, pp. 3-43, March, 1983.

2. Lubomir Bic and Alan C. Shaw, *The Logical Design of Operating Systems,* Prentice-Hall, Englewood Cliffs, NJ, 1988.

3. E. J. Coffman and P. J. Denning, *Operating System Theory,* Prentice-Hall, Englewood Cliffs, NJ, 1973.

4. Jaques Cohen, ''Non-Deterministic Algorithms,'' *Computing Surveys*, vol. 1, no. 2, pp. 79-94, June, 1979.

5. M. J. Flynn, *Very high-speed computing systems,* Proceedings of the IEEE, 1966, pp. 1901-1909, IEEE, New York, 1966.

6. C. A. R. Hoare, *Communicating Sequential Processes,* Prentice-Hall, Englewood Cliffs, NJ, 1985.

7. INMOS, Corp., *OCCAM Programming Language,* Prentice-Hall, Englewood Cliffs, NJ.

8. Robin Milner, *A Calculus of Communicating Systems,* Lecture Notes in Computer Science, number 92, Springer-Verlag, Berlin, 1980.

9. Terrance W. Pratt, *Programming Languages: Design and Implementation,* Prentice-Hall, Inc., Englewood Cliffs, NJ, 1984.

10. S. S. Reddi and E. A. Feustel, ''A Conceptual Framework for Computer Architecture,'' *Computing Surveys*, vol. 8, no. 2, June, 1976.

11. Daniel P. Siewiorek, C. Gordon Bell, and Allen Newell, *Computer Structures: Principles and Examples,* McGraw-Hill Book Company, New York, 1981.

12. D. E. Stevenson, *Considerations in System Design for Scientific Supercomputer Applications,* Proceedings, Third International Conference on Supercomputing, International Supercomputing Institute, Inc., St. Petersburg, FL.

High Performance Parallel Computers for Science: New Developments at the Fermilab Advanced Computer Program*

T. Nash, H. Areti, R. Atac, J. Biel, A. Cook, J. Deppe, M. Edel, M. Fischler, I. Gaines, R.Hance, D. Husby, M. Isely, E. Miranda, M. Miranda[a], T. Pham, and T. Zmuda

Advanced Computer Program
Fermi National Accelerator Laboratory†
Batavia, IL 60510 USA

E. Eichten, G. Hockney, P. Mackenzie, H. B. Thacker and D. Toussaint°

Theoretical Physics Group
Fermi National Accelerator Laboratory+
Batavia, IL 60510 USA

ABSTRACT

Fermilab's Advanced Computer Program (ACP) has been developing highly cost effective, yet practical, parallel computers for high energy physics since 1984. The ACP's latest developments are proceeding in two directions. A Second Generation ACP Multiprocessor System for experiments will include $3500 RISC processors each with performance over 15 VAX MIPS. To support such high performance, the new system allows parallel I/O, parallel interprocess communication, and parallel host processes. The ACP Multi-Array Processor, has been developed for theoretical physics. Each $4000 node is a FORTRAN or C programmable pipelined 20 MFlops (peak), 10 MByte single board computer. These are plugged into a 16 port crossbar switch crate which handles both inter and intra crate communication. The crates are connected in a hypercube. Site oriented applications like lattice gauge theory are supported by system software called CANOPY, which makes the hardware virtually transparent to users. A 256 node, 5 GFlop, system is under construction.

* Talk given by Thomas Nash at the Workshop on *Computational Atomic and Nuclear Physics at One Gigaflop*, Oak Ridge, TN, April 14-16, 1988.

† Fermilab is operated by Universities Research Association, Inc. under contract with the U.S. Department of Energy.

1. INTRODUCTION

High energy physicists, like most scientists, have always wanted more computer power than they could afford to buy.[1] In the commercial marketplace, the emphasis is on software backward compatibility, product differentiation, and isolation of a client "herd". On the scientist's computing agenda, raw processing power in a relatively easy to use form, not corporate profitability, is the dominant issue.

Despite the fact that industry is not motivated by the rest of the market to provide the extremely cost effective computer *systems* demanded by much of science, it does provide an extraordinary array of *components* (chips, modules, peripherals, work stations, software, etc.) that can be assembled into what science requires. For over 5 years the Advanced Computer Program (ACP) at Fermilab has drawn from industrial components to design and produce usable parallel computer systems of such cost effectiveness that high energy physics (HEP) experiments are now being carried out that would otherwise be unthinkable.

In addition to its dependence on the latest offerings of industry, the ACP also takes advantage of developments coming from computer science. However, just as with industry, the motivations of natural and computer scientists have significant areas of divergence. When commissioning a new architecture prototype the interest of computer science primarily extends to proof of principle. Although also concerned with the latest from computer science, the ACP has the goal driven motivations of its scientific clientele. The ACP's activities, therefore, fall in the rather wide gap between the exploratory spirit of computer science and the production reliability and conservatism of industry. ACP systems must both take advantage of advanced concepts and components and be robust enough to satisfy a large community of demanding users.

In common with computer science, the ACP is aiming toward reasonably general purpose parallel computers. The difference is that the ACP cannot focus, as computer science usually does, on the often academic complications of the end goal. Rather, the ACP must take a step wise approach, dealing with the easier, more obviously parallel, problems first. At each step as more complex issues are addressed, there must be a fully working, commercializable machine that delivers the maximum cost effectiveness available at the time.

With their different perspective, natural scientists have brought to the computer development community new philosophical and architectural approaches to investigate. The dynamic nature of scientific problems means that programs are frequently changed, often extensively. The scientific user generally has a deep understanding of the structure of the problem and the capability and willingness to map the algorithm to optimally cost effective computer architectures. Issues of software compatibility can certainly become painfully acute at mundane levels like compiler syntax. However, when it comes to the global scale of a problem, scientists are virtually unique at this time in their receptiveness to explicit (user directed), as opposed to implicit (automatic), decomposition of programs onto parallel computers.

Fox and Seitz at Cal Tech, driven initially by the needs of theoretical physics, established the viability of hypercubes, the quintessential explicitly parallel, local memory architecture.[2] These are now the subject of much attention in computer science which earlier had focussed heavily on global memory architectures and attempts at automatic parallelization of whole algorithms. Similarly, the natural parallelism inherent in the experimental HEP computing problem (running the identical reconstruction program on many millions of different events) suggests a simple parallel processing solution. The pioneering emulator work by Kunz *et al.* at SLAC demonstrated the feasibility of using multiprocessor systems to provide cost effective computing for the reconstruction of experiment events.[3]

Almost a decade after the first SLAC emulator, powerful 32-bit microprocessors allowed the ACP at Fermilab to develop even more convenient and cost effective event oriented parallel processing systems using many more CPUs.[4] Such systems are now an acknowledged important component of computing in high energy physics. They have made crucial contributions to data analysis in a number of current experiments, and will be indispensable for the large experiments of the SSC era. The first generation ACP systems are rapidly becoming a standard. There are over 30 installations in universities and laboratories worldwide, primarily, though not exclusively, for HEP applications. The first of these ACP systems, which provide CPU power at a cost of less than $2500 per VAX 780 equivalent, were brought on line two years ago. The initial 110 processor system, along with several newer installations, have been heavily used at Fermilab for physics event reconstruction.

This paper will describe new work from the ACP in two areas: a second generation multiprocessor targeted at experiment applications, and a multi array processor "supercomputer" for the site oriented problems of theoretical physics, principally lattice gauge calculations. The mandate for ACP development remains the computing needs of high energy physics. However, the potential applicability of these new systems, particularly the one aimed at theoretical physics, is far broader than HEP alone.

We will first cover the new ACP systems for experimental HEP which, early in the next year, will provide well over an order of magnitude increase in cost effectiveness over the original systems. This second generation project will also allow much higher bandwidth for both I/O and interprocessor communication, and will have software tools allowing almost any UNIX, VMS, or (potentially) VM based processor to be used equivalently as a node or "front end" in a multiprocessor ACP system. Perhaps most important is the way in which this new system allows for integration of powerful "back end" multiprocessors, now usable for both reconstruction and physics analysis, into a modern Ethernet based workstation environment.

The multi array processor system for theory will be described next. Programmable in C and FORTRAN and supported by a user interface that makes transparent the details of the hardware architecture, this system is also at the highest levels of cost effectiveness. A 5 GigaFlop (peak), 2.5 GByte, 256 node, system is expected to be running in early 1989 at a cost of approximately $1 million. This system, like all others from the ACP, is intended for quick commercialization so that it will be available to other institutions with needs similar to Fermilab. At the end of the paper, we will outline expectations of future directions ACP developments may take including the likely merger of the experimental and theoretical machines.

2. THE SECOND GENERATION ACP MULTIPROCESSOR

The continued saturation of computers (including ACP systems) by HEP experimenters motivates the development of a second generation of the successful ACP Multiprocessor system. Industry is continuing to provide a stream of often surprisingly powerful technology that we can harness to meet the severe challenges of this science. Frankly, some of industry's new components are making our task particularly pleasurable and rewarding these days. A variety of new and increasingly powerful microprocessors are now available to incorporate into multiprocessors. Beyond the new versions of existing processors (like the 25 MHz 68030s replacing 16 MHz 68020s), there are entirely new families of chips. Most of these are based on the Reduced Instruction Set Computer (RISC) architectural philosophy. It was noted in the 1970s that many complex instructions in traditional machines with large instruction sets, like the IBM 360s and DEC

VAXes, were rarely used. They effectively increased the cycle time for all instructions because they mandated extensive microcode. RISC machines are generally pipelined with no microcode and very fast instruction cycles. They defer complex instructions to software.

New processors that will offer at least a factor of three, and in one case as much as a factor of twenty, more performance than the first generation ACP processors based on 68020s include: the MIPS R2000 and R3000 RISC chips; the Fairchild, now Intergraph, Clipper chip set; the AMD 29000 RISC chip; the Sun SPARC RISC chip; the INMOS T800 transputer; the Motorola 88000 RISC chip; the National Semiconductor 32532 processor. Unlike earlier processors, many of these have usable FORTRAN and C COMPILERS and UNIX operating systems even before the hardware is available.

The broad availability of UNIX is particularly important. As we will describe below, the new ACP architecture will support any processor running UNIX (or VMS) that can be connected via VME or Ethernet. It is very difficult to predict with any certainty which processor or commercial single board (or single slot) computer (SBC) will be most cost effective in the future. The openness of the ACP system permits competitive purchase of processor nodes based on performance benchmarks and price.

The large increase in CPU power available for the Second Generation ACP System requires a redesign of the multiprocessor hardware and software system architecture to remove bottlenecks in current systems that would be felt at the higher performance levels. The bottlenecks are I/O and interprocessor communication bandwidth, and the CPU power available for the host process. Continuing the successful strategy of attacking computing limitations with parallelism, the new architecture solves I/O and host limitations by supporting parallel I/O and parallel host processing. Moreover, to avoid mini computer bus bandwidth restrictions, any node may take on host functions including I/O through controllers in its own local crate. Though significant, the changes in the new system hardware and software are designed to be as transparent as possible to users of the original system. In the following, we describe the fast new ACP SBC, and then the new interconnection modules and topologies, the new (parallel) I/O options, and the new software system architecture that are all designed to support the large increase in performance that will be available.

2.1. A Fast New ACP Processor Node

As it has in the past, the ACP is encouraging competition in SBCs targeted at parallel processing by developing an extremely cost effective VME SBC. The choice of the R3000 RISC processor from MIPS Computer Systems, Inc. for this design was based on performance evaluations of the microprocessors referred to earlier. The ultimate standard is how real physics code runs in a high level language, since it does not matter how many million instructions per second (MIPS) the CPU can execute if the instructions are not useful in FORTRAN or C and if the compilers fail to provide sufficient optimization. The ACP has performed benchmarks on several of the new chips using a suite of high energy physics FORTRAN programs (a small Monte Carlo/track reconstruction package; an actual fixed target tracking code running on experimental data; and a floating point intensive theoretical calculation). These benchmarks were run on a 16 MHz MIPS R2000 system. For the three benchmarks, performances were 7.9, 6.4, and 7.4 times that of a VAX 780, which is generally accepted as the standard 1 MIP normalization. The new ACP node will use an R3000, the new improved version of the R2000. Carefully taking into account processor, cache, as well as clock speed differences, leads to a projected HEP code performance of 12-15 VAXes for 25 MHz R3000 boards and 16-20 VAXes for the 33 MHz versions we expect to use.

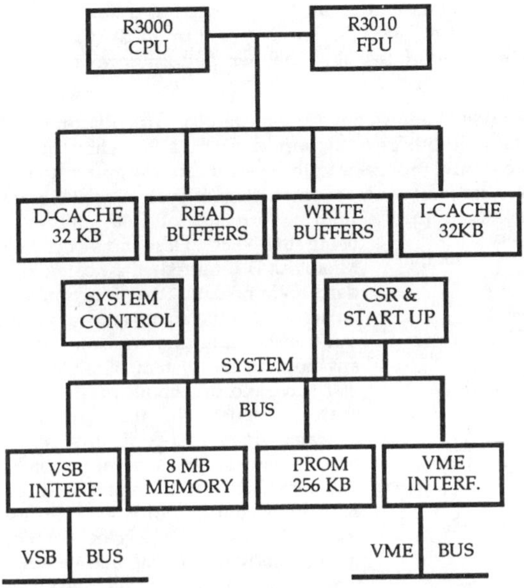

Figure 1. The ACP MIPS processor, block diagram.

The ACP MIPS processor module (Figure 1) consists of:

1) a 25 or 33 MHz MIPS R3000 CPU;
2) a 25 or 33 MHz MIPS R3010 floating point unit;
3) Four Write Buffers;
4) a 32 KByte instruction cache;
5) a 32 KByte data cache;
6) 8 MBytes of parity checking main memory, made up of 1 Mbit (100 nsec nibble mode access) DRAMs, expandable to 24 MBytes via VSB (VME System Bus);
7) interval timers that can interrupt the CPU;
8) a full function VME Master Slave interface supporting 20 MBytes per sec block transfers;
9) a full function VSB interface for peripherals and memory extensions.

The ACPMIPS processor module will be packaged as two connected boards that will fit in one standard VME slot. One board will contain the processors and cache; the other board will have main memory and the VME and VSB interfaces. Up to two extra memory boards may be plugged into other slots and connected via VSB to allow a total of 24 MBytes. The modular nature of this design means that as new components (denser memory or faster processors) become available, only one board at a time needs updating. With 4 Mbit DRAM, expected in a year or so, 32 MBytes will be possible without an expansion board (or up to 96 MBytes with maximum expansion).

The new CPU module will provide high level language processing power with a cost effectiveness of well under $200/VAX 780 equivalent, based on the physics benchmarks and features listed above. The FORTRAN compiler is the best we have encountered for a microprocessor. It supports VMS extensions and compares favorably with the VMS compiler in convenience and sophistication. Since the MIPS CPU chip has on chip memory management, the board will be able to run the full UNIX operating system, booting either from a VME disk drive or using the Network File System (NFS) over the Branchbus. Full UNIX program development tools are available. This processor will form the cornerstone of the second generation ACP systems.

2.2. New Interconnect Topologies

The original ACP multiprocessor used a single (MicroVAX) host which was the master of large numbers of microprocessor nodes. This system has been extensively described in earlier papers.[4] Here we discuss only the significant new improvements. The ACP Branchbus was developed to link several high performance commercial local bus crates (like VME) to a host and/or a data acquisition system. It is optimized for high speed block transfers.[5] No commercial alternative was available. The original Branchbus is a 32-bit bus connecting a single master (QBus, Unibus or FASTBUS) to multiple crates (VME), with block transfers at up to 20 MBytes/sec. Improvements to the Branchbus system of interfaces allow higher performance and more complex interconnection schemes.

The new VME Branchbus Controller (VBBC) allows any VME master, in particular any node (or smart I/O device controller) in an ACP multiprocessor system, to act as a Branchbus master and read or write to any Branchbus address. Any processor in the system can communicate with any other processor without host intervention, allowing the more elegant system architectures described below. Figure 2 shows a block diagram of a high performance off-line ACP system using the VBBC. Here no specialized MicroVAX is required as a host; a standard node takes on that function, and an accompanying smart tape controller can transfer data directly from tape to any node in the system. The VBBC is a VME slave and Branchbus master. It is a shared resource in the VME crate, allocated on a first-come first-served basis by a test-and-set bit in its control register. Once programmed with the Branchbus control words and address, the Branchbus cycles occur transparently to the VME master which simply reads or writes data to the VBBC. First prototypes of the VBBC are now working.

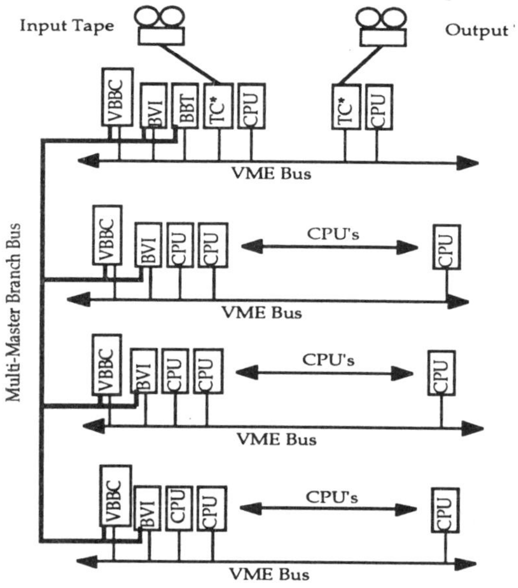

Figure 2. A High Performance Off-Line ACP System

As implied above, the Branchbus now supports multiple masters. A distributed arbitration scheme similar to that used on the SCSI bus allows up to 16 masters to share the bus. Existing masters (Q/UBBC modules) can be used in multi-master systems by replacing their existing Branchbus Interface Daughter Board with a new Multi-Master Branchbus Interface Daughter Board which handles the arbitration transparently to the user.

The ACP Branchbus Switch allows full crossbar interconnection of up to 16 Branchbuses (or more using multiple switches). With this switch, any Branchbus master device can connect to any slave in the entire switch connected system. All channels of the switch can be active simultaneously. For example, eight of the Branchbuses could be connected to the other eight, all transferring data simultaneously giving an aggregate bandwidth of 8 X 20 MBytes/sec or 160 MBytes/sec (in addition to any local bus activity on any of the VME crates in the system). Thirteen Texas Instrument 16 X 16 X 4 bit crossbar chips (TI 74AS8840)are used for the main switching elements. The Switch is a backplane in a 6U by 280 mm Eurocard crate. Modules may be plugged into the Switch Crate (see Figure 3) much as with VME. However, instead of the signals being connected in a bus structure, each slot in the crate is a crossbar switch point. The first two Switch crates are built and working.

Two modules now exist that plug into the Switch Crate. One is the Branchbus Switch Interface Board (BSIB) which converts the differential RS485 signals on the standard Branchbus cables to the single ended TTL version of Branchbus used on the Switch backplane. The BSIB brings one standard Branchbus, with its VME crates, host, Fastbus, etc., into a port of the Switch, and allows them to be switched to whatever else is plugged into the Switch Crate, such as other Branchbus circuits. Use of the Switch in this way will allow multiprocessor systems to obtain extremely high bandwidth for interprocessor communication. The second existing module that plugs into the Switch is the Floating Point Array Processor (FPAP), a 20 MFlop device, that is used for theoretical physics, primarily lattice gauge, calculations. The FPAP is described later in detail, as is the innovative "better than a hypercube" architecture that the Switch allows. Important future applications of the versatile Switch are described in the Section 4.

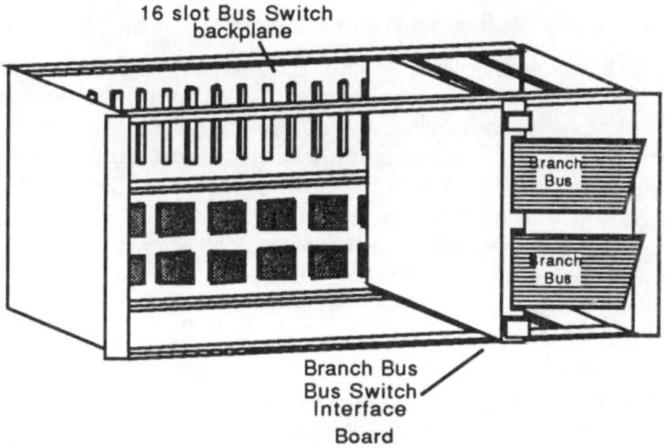

Figure 3. The ACP Branchbus Switch. The backplane uses single ended TTL Branchbus protocol.

2.3. Parallel Input/Output

Although it will continue to be possible to read and write tapes through a VAX or MicroVAX into an ACP system, high performance operation will take advantage of I/O devices that interface directly to the multiprocessor system bus. Unlike Unibus and QBus tape drives, VME interfaces will give a bandwidth potential, depending on the I/O device, of up to the 20-30 MBytes/sec allowed by VME rather than the roughly 1 MByte/sec possible with minicomputer buses. One such VME tape interface, the Ciprico TM3000, has been used for data acquisition at Fermilab.

Parallel I/O is particularly encouraged by the availability of inexpensive, high capacity mass storage devices which interface directly to VME. It is made possible by the availability of I/O directly in the node crates and by the new capability of a processor and/or its intelligent I/O controller to write through a VBBC from one crate to another. The multiprocessor system will have available to it many devices all reading and writing simultaneously, allowing the total I/O bandwidth to be increased to whatever level is required.

New I/O devices are replacing standard 6250 bpi magnetic tape. These are very appealing to HEP experiments anticipating huge amounts of data, such as one approved Fermilab experiment which is planning to record tens of billions of events. At this time video tape cassettes appear most promising. The 8 mm format can pack well over 100 times the data in a given volume of shelf space as can conventional tapes, and they are at least an order of magnitude more cost effective. Current devices allow bandwidths comparable to those achieved with standard 6250 bpi tapes, with further improvements expected in the next year.

The ACP has tested the 8mm video tape device manufactured by Exabyte Corporation. Currently available devices store 2 Gbytes (as much as 12 conventional tapes) on a standard $10 tape cassette. The drives cost less than $3000 and can deliver data at 250-KBytes/sec through a SCSI bus interface to VME or QBus. Both density and speed are expected to double in the next year.

The Exabyte drive package allows the design of a cheap and simple mechanical loader. Such a design is underway. It will incorporate a reader for optical bar code labels to insure mounting of the correct set of tapes from a large data sample. Automating tape loads eliminates the time for human operators to mount tapes which becomes significant when tapes are being processed at high speed.

The ACP plan at this writing is to support the Exabyte drives. However, it is important to emphasize that no standard has yet emerged in this rapidly developing field, and the long-term reliability of such devices has not been established. Nonetheless, it is clear that one can count on far better I/O performance than has been available.

2.4. Software System Architecture

The redesign of the system software for the Second Generation ACP System will support the greatly improved performance and flexibility of the new processing, I/O, and communication hardware, while reducing the complexity encountered by both beginning and sophisticated users.[6] It will allow existing applications to run with minimal changes, yet will provide a variety of powerful new features to allow users to realize the full potential of the new processors. Integration will be possible into a variety of computing situations, including large mainframe computer centers and the traditional VAX or MicroVAX host. Most important, in our view, is a distributed computing UNIX (or VMS) workstation environment in which the multiprocessor will function as a fully integrated back end engine directly controlled from the workstations. The general hardware environment supported is shown in Figure 4. Because of the emphasis on portable, nearly universal standards, it will be straightforward to incorporate any computer that runs UNIX and/or can communicate via TCP/IP over Ethernet. This makes the system open and receptive to future requirements and product options.

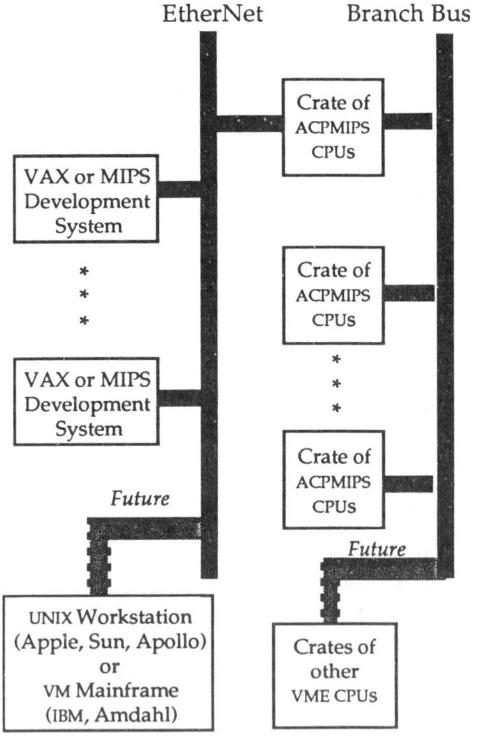

Figure 4. Second GenerationACP System: General hardware and network environment.

As before, the ACP System Software is a tool to make it easy for physicists to bring their programs to a high performance multiprocessor environment. An application for the second generation system is decomposed into a set of cooperating processes. Support for running these sets of processes as a job, including interprocess communication and synchronization, startup, etc., is provided by the ACP System Software. The processes run on an ACP Multiprocessor System interconnected by Branchbus and on any associated UNIX or VMS (later possibly also VM) computers connected via Ethernet and TCP/IP. They may be distributed as appropriate over the available computers. In this way a multi process job may be tested first in a single machine and then with increasing numbers of nodes. Program development is done using the full set of UNIX (or VMS) tools, including compilers, linkers and debuggers, of the computer on which the process will run.

Any node process can assume the functions previously exercised only by the VAX host processor. Any node process in the system can do I/O, reading or writing data tapes and accessing disk files. And any node process in the system can do `send` or `get` operations to or from an individual process (chosen by the system software from a class or rank of node processes) or set of processes in a given class or rank. As before, the system software will automatically find an available node process for the user. The ability to `send` and `get` to multiple nodes in a class allows `broadcast` and `accumulate` type operations.

Along with the traditional `send` and `get` type of ACP communication routines, there will be a variety of more primitive, yet powerful and easy to use, interprocess communication mechanisms. A process may send a block of data directly to a block of virtual address space in another process, or it may call a subroutine in another process (remote subroutine call), or it may send a small data packet (a message) to another process. Users of the new system will have direct access to process queues which they may define as they require or use in standard, traditional ACP defined ways (like node process *ready* or *complete*). Synch points provide a way for all processes in a class to synchronize program execution.

There are many possible process configurations that the new system can support. An example for a reconstruction problem with multiple input tapes is shown in Figure 5. Note that this is a software configuration; the actual hardware connection of the nodes is over Branchbus via VBBCs and Bus Switches (if necessary) and is transparent to the programmer. Nodes in the top rank read events from data tapes and pass them along to either class 2 or class 3 nodes, which process events of different trigger types. Nodes in the bottom rank collect events from any nodes in the middle rank, either class 2 or class 3, for output to tape. Another important configuration has enormous implications for physics data analysis (as opposed to reconstruction). Here, the top rank of nodes is the same as in Figure 5, and the second rank consists only of one or more workstations and, perhaps, a single data recording process. With such a configuration, a whole experiment's data base of data summary tapes (DSTs) can be read and analyzed in parallel in much less than an hour. The traditional means of passing hundreds of DSTs through a computer center for a physics analysis pass often takes weeks.

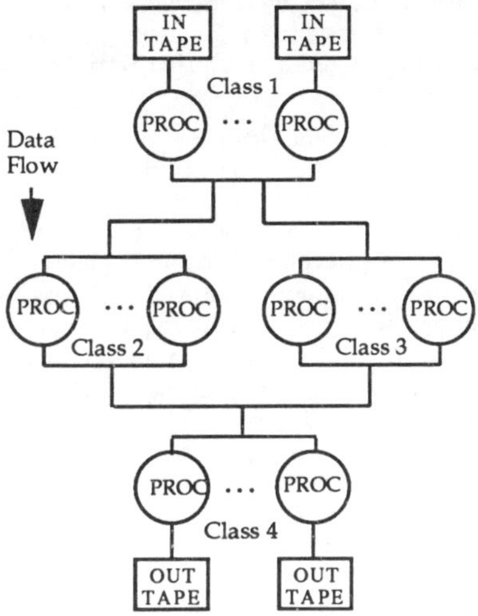

Figure 5. A multi rank configuration. Processes are indicated by circles.

To summarize, the Second Generation ACP System architecture is designed to support large increases in processing power and in I/O and communication bandwidth. It provides a set of hardware and software building blocks so that a system can be matched to the set of applications it will run. One can provide enough I/O devices and Branchbus interconnects so there are no bandwidth limitations. The number of basic CPU nodes is determined by processing power requirements, while special computer nodes, such as workstations or mainframes, can be incorporated into the Ethernet side of the system when needed and available. As each job is run on the system, nodes are assigned to run particular user processes (input, output, event processing, etc.) as appropriate. Not only the traditional compute bound event reconstruction, but also more I/O intensive data analysis will find a home on these systems.

3. THE ACP MULTI-ARRAY PROCESSOR SYSTEM FOR THEORISTS

The ACP Multi-Array Processor System (ACPMAPS) is a highly cost effective, local memory parallel computer designed for floating point intensive grid based problems. The project is a joint effort of the ACP and Fermilab's Theoretical Physics Group. Processing nodes of the system are single board array processors based on the FORTRAN and C programmable Weitek XL chip set. They are connected by a network of very high bandwidth 16 port crossbar switches. The architecture is designed to achieve the highest possible cost effectiveness while maintaining a high level of programmability. At Fermilab the primary application of the machine will be lattice gauge theory.

To obtain some estimates of the computing needs of lattice quantum chromodynamics (QCD) one can consider the calculation of the deconfining temperature in SU(3) gauge theory without quarks,[7] which is one of the most solid four dimensional calculations done so far. Something like 500,000 MFlops - hours (peak, ~70% delivered) were used on a Star ST-100 array processor. This calculation required a lattice spacing of less than 0.1 fermi and a volume of close to (2 fermi)^3, resulting in lattices with spatial sizes of up to 19^3. It is virtually certain that calculations with quarks will require even larger lattices than this for comparable accuracy. Lattices with space-time sizes 32^4 to 64^4, requiring 1-20 GBytes of data memory, are a reasonable guess. Calculations of hadron masses in the approximation of ignoring dynamical quarks have not yet achieved a reasonable understanding of calculation errors, even on Cray-sized supercomputers. Although algorithms for the inclusion of dynamical quark effects have made tremendous progress in the last few years, at present they still seem to require at least two orders of magnitude more computer time than comparable calculations without quarks. It is thus clear that large increases in combined CPU power *and* algorithmic power are still required even for simple QCD calculations.

The aim for the new ACPMAPS machine is to deliver such large amounts of memory and CPU power at the lowest possible cost, without compromising the programmability required for rapid algorithm development which is just as important as raw computing power in achieving the goals of lattice gauge and other problems in theory. Descriptions of the other large-scale computer projects aimed at lattice gauge theory may be found in the references[2,8].

A 16 node system is being built this summer. Two switch prototypes are working and tested. Four Floating Point Array Processor (FPAP) node modules are also working and undergoing rigorous testing. They have successfully run extensive physics code. Fermilab intends to proceed to a 256 node (5 GFlop for about $1 million) system as soon as the 16 node system is operational. Parts are being procured for this system which will be assembled at the end of the year. Given the communications bandwidth noted below and the large amount of memory per node, it has been calculated that performance for appropriate lattice gauge algorithms should increase linearly well past 256 nodes. Maximum system size is 2048 nodes. The system is being designed in the ACP tradition to be commercialized and available to other institutions.

3.1. Architecture Overview

A block diagram of the system is shown in Figure 6. The individual single board FPAPs have peak performance of 20 MFlops. Performance of key kernels (SU(3) multiplies) have been measured on the prototypes to exceed 15 MFlops/node. For lattice gauge physics, a performance standard is the link update time. Using the Kennedy-Pendleton heat bath algorithm (pure gauge) a single FPAP has been clocked at under 800 μsec per link update. Depending on the algorithm, this corresponds to a real performance of 4-10 MFlops/node. Each FPAP contains 8 MBytes of data and 2 MBytes of program memory. (In the event of memory shortages, the FPAPs can be configured with a minimum of 4 MBytes for data.)

The FPAPs are plugged into a crate whose backplane is a 16 fold bidirectional high speed crossbar. This is the Branchbus Switch Crate described earlier. The nodes can speak with each other in pairs at a full 20 MBytes/sec simultaneously. The architecture of ACPMAPS is a hypercube network of such crossbar switch crates each supporting 8-16 FPAPs. In a typical configuration 8 array processor nodes will be plugged into each switch crate along with up to 8 BSIB I/O modules (also described in an earlier section) that interconnect crates in a hypercube (or better, if extra interconnects are desired).

Processing nodes do not participate in any communication activity other than their own. This is an important distinction from traditional hypercube implementations. The switches handle intra and intercrate routing automatically. The system therefore does not operate with all

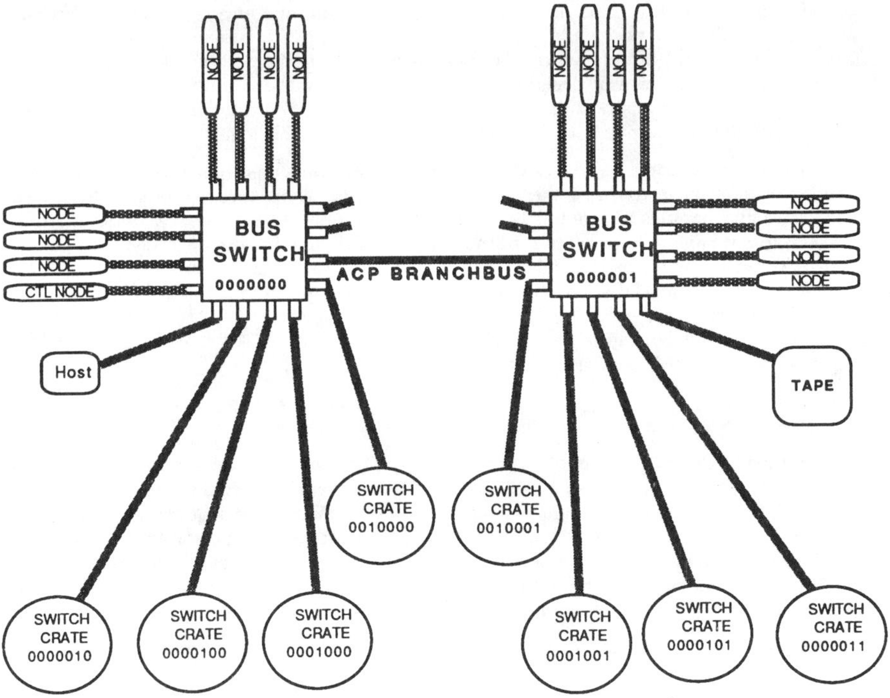

Figure 6. ACP Multi Array Processor System: 256 node configuration.

node programs (and/or communications) in lock step like an SIMD machine, as is the case in most of the other projects of this type[8] (Columbia, IBM GF11 and APE). It also does not strongly favor local communication (as existing hypercubes[2] do). It thus allows for any conceivable new lattice algorithm unconstrained by synchronous or local communication requirements. Despite its algorithmic flexibility the system ranks as the best (or nearly so, we won't argue) in terms of cost effectiveness of MFlops/$.

In addition to the flexible global architecture of the system, there are two important aspects of the FPAP itself that distinguish it from the CPUs of the other more special purpose lattice gauge processors. First, the FPAP memory is neither too large nor too small, but is ideally matched to the FPAP performance and communication capabilities and the demands of lattice gauge (and presumably other site oriented) problems. Secondly, the FPAP sticks closely to the Weitek XL architecture, using only one floating point processor per board. This allows the use of C and FORTRAN compilers and greatly simplifies the microcode.

The asynchronous communication and MIMD (multiple instruction, multiple data) processing architecture, in distinction to the more common synchronous communication, SIMD (single instruction, multiple data) approach, is one of the most important features of the system. There are many advantages to this type of architecture. It is very flexible: it can handle problems which are awkward or impossible in synchronous SIMD such as, in the case of lattice gauge, heat bath and incomplete LU decomposition algorithms and random lattice problems. The allowed sizes and shapes of the lattices are independent of the details of the hardware. The node structure of the machine can be made invisible in much or all of the high-level user code, resulting in improved programmability. This also results in improved fault tolerance, since the system can be reconfigured readily if a node fails, without requiring changes in user software or allocating nodes as spares. Complications which have to be faced include the potential for synchronization conflicts. This requires care in designing and understanding the communications system. In addition, a nontrivial system software design effort is required to ensure that overheads associated with the communications software are kept to acceptable levels.

A major new package of software (CANOPY) has been developed for this system. Theorist users need think only in terms of sites and fields on sites. The system automatically allocates sites to nodes and handles all site to site communication whether on the same node or another. Thus users do not have to know details of the hardware for effective use of the system. Routines that are used heavily will be microcoded. The skeleton of all applications are written in FORTRAN or C using a series of special subroutine calls that make the programs particularly readable for lattice gauge theorists and others with site oriented algorithms. In this way, despite ease of use and flexibility, the system can approach 10 MFlops/$4000 node in FORTRAN or C.* The CANOPY system software is described in more detail below.

3.2. The Floating Point Array Processor Module, Communication, and I/O

The initial ACPMAPS FPAP nodes are single board floating point array processors using the Weitek XL chip set which contains a 32 bit, 20 MFlop (peak) floating point unit, an integer processor, 32 floating point and 32 integer registers, and an instruction sequencer. The chip set as a whole is programmable in FORTRAN and C, at some sacrifice in performance. Thus, these modules incorporate the functions of a high level language programmable single board computer and a high performance floating point array processor. No external CPU is required as a controller for these stand alone floating point engines.

The FPAP modules (Figure 7) contain the XL chip set, the data and code memory, and the

* Memory prices are fluctuating widely at this writing. Costs given here are based on the (high) summer 1988 DRAM prices.

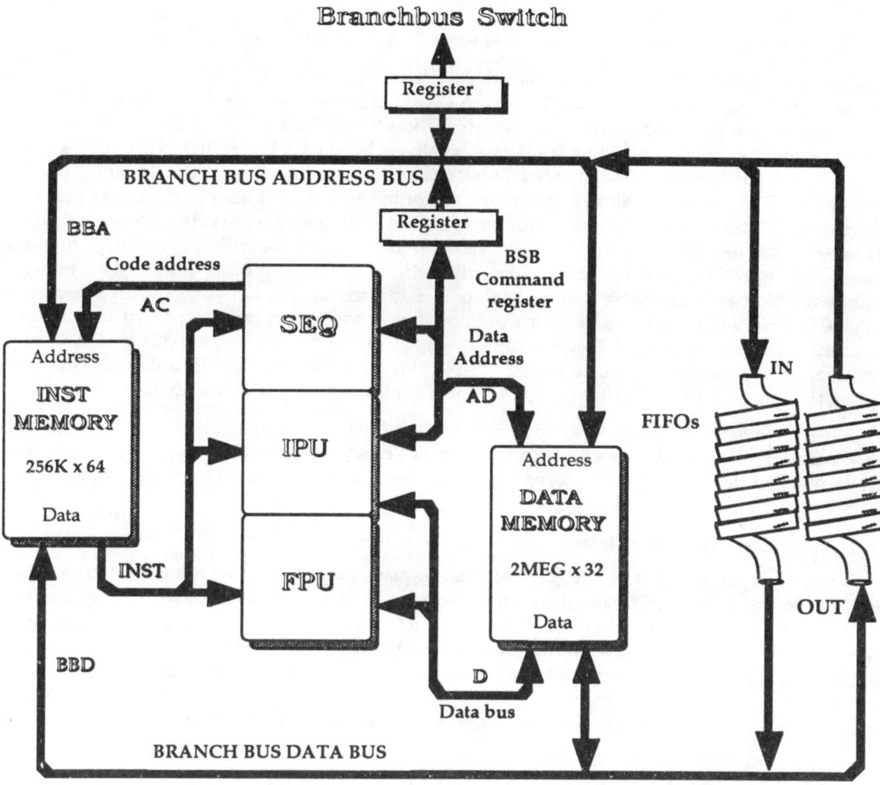

Figure 7. Schematic Design of the Floating Point Array Processor (FPAP)

interface logic and input and output queues for communicating with the crossbar switch crates. One floating point unit is used per node, in contrast to the designs of most of the other machines aimed at lattice gauge theory. In addition to being a flexible and sensible design for a wide variety of problems, this was dictated by the desire to be able to use the Weitek FORTRAN and C compilers for the XL chip set.

The 2 MBytes of program memory and 8 MBytes of data memory is made from 1 Mbit 80 nsec access time page (or static column) mode dynamic RAM chips. In page (or static column) mode these memory systems can deliver data at a rate of one word per 100 nsec. This rate is fast enough that little additional efficiency would be gained in most lattice algorithms by replacing some of the DRAM by faster, more expensive static RAM. The memory chips constitute at least a third of the total cost. The memory to power ratio provided (8 MBytes to 20 MFlops) is larger than that provided by most other machines of this type, and is larger than is required by presently existing algorithms for simulating full QCD including internal fermion loops. It is approximately appropriate for calculations in the valence approximation, ignoring fermion loops. Algorithmic improvements over the next few years will certainly change the required ratio. It seems likely that the possibilities which will increase the required ratio (preconditioning and Fourier acceleration of quark propagator calculation, Fourier acceleration of gauge simulation) are currently more promising than those which reduce the amount of memory required per CPU cycle (such as adding nonlocal operators to the action to reduce finite lattice spacing errors) and that the large amount of memory could easily become crucial in the years to come.

The nodes are plugged into a network of Branchbus Switch Crates (described in Section 2.2) whose backplanes handle full sixteen port crossbar switching at bandwidths of 20 MBytes/second per connection. This yields a total bandwidth of 2.56 GBytes/sec for a 256 node machine. A cluster of 8-12 nodes is attached to each switch. The switches are connected in a hypercube, which may be augmented by additional communication channels along heavily used paths. This structure allows the nodes to communicate as if they were connected in a conventional hypercube arrangement, but more than this, it allows any node to communicate at full speed with *any* other node, allowing efficient running of algorithms requiring nonlocal communications. The Switch Crates allow any node to access any other node's data memory without needing to know where the other node is located on the network. With the current Switch Crate hardware, systems of up to 2048 nodes are possible before this transparent nonlocal communication feature is lost. The Branchbus Switch Crates will also be used in a variety of high performance experimental particle physics applications of the ACP Multiprocessor System, including the Level-3 programmable trigger for the Collider Detector at Fermilab, CDF.

The Exabyte video technology tape drives, described in an earlier section, will be used for check pointing long calculations and for archiving of gauge fields and propagators. One drive will be attached to every switch crate, enabling all of memory to be stored in under five minutes.

3.3. CANOPY System Software

Lattice gauge theories are part of a large class of grid-based problems derived from discretization of a set of differential equations which are very suitable for a parallel architecture like this one. The natural breakdown of the problem is to assign a certain subset of the sites in the space or spacetime to each node, which stores the data for the field variables defined on the sites assigned to it in its local memory and does calculations for its sites. The system software[9], known as CANOPY, has been designed to shield the user as much as possible from the hardware dependent node structure of the parallel architecture. The user thinks in terms of sites not nodes.

User programs are divided conceptually into two pieces: the control program, which is called from a MicroVAX host or mainframe VAX and runs on the control node, and site subroutines, which run on the individual nodes. The control program manages the execution of lattice-wide tasks. It is typically written in ordinary FORTRAN or C augmented by a set of system subroutines for dealing with global concepts (e.g., field memory, lattice wide tasks) which are distributed over all the nodes and require special treatment. The beginning of the control program includes statements like the following:

```
call define_periodic_lattice ( ndims, sizes, lat1 )
call define_field ( lat1, quarksize, q )
call define_field ( lat1, quarksize, q1 )
call complete_definitions
```

The routine `define_periodic_lattice` tells the system that our problem contains one lattice called `lat1` of `ndims` dimensions with the size of each dimension contained in the array `sizes` and with standard hypercube connectivity. More general user defined connectivity are allowed. It is possible to define several lattices in the same program for block spin renormalization group or multigrid algorithms. The routine `define_field` tells the system that memory will be required for storing two fields identified by `q` and `q1`, each with `quarksize` components for each site of `lat1`. The routine `complete_definitions` calls routines which assign specific sites to specific nodes, allocate memory in the nodes for the field data and site structures, and set up structures for each site pointing to the memory areas of adjacent sites of the lattice.

A control node subroutine which operates on a field `q` with an operator `dslash_` and stores the result in another field `q1` would be written as follows.

```
subroutine dslash ( q, q1 )
        .
        .
        .
call do_task ( dslash_, lat1,
        pass$, q, 1,
        pass$, q1,1,
        end$)
return
end
```

The system subroutine `do_task` passes to all the nodes a pointer to a subroutine `dslash_` which operates on a single site and a pointer to a list of sites on which to operate, which may be the entire lattice `lat1` or some previously defined set of sites such as `red_sites`. A system routine on the node, invisible to the user, calls `dslash_` for all the sites in the set of sites which have been assigned to the node. `do_task` may also be used to pass (`pass$`) to the nodes parameters required by the site subroutine (like the field identifiers `q` and `q1`) and to integrate (`integrate$`) data returned from the individual nodes.

The site subroutines access and replace data from global fields with subroutines like:

```
call get_field   ( q,   site, qtemp )
call put_field   ( q1,  site, qtemp )
```

which determine if the desired data is already resident on the node and open a channel to the communications hardware if necessary.

Most site subroutines can be written in FORTRAN or C. CPU intensive kernels such as SU(3) matrix multiplication and essential routines like `dslash_` will be microcoded for maximum efficiency. We expect that lattice gauge algorithms prepared in this way will run at up to 10 MFlops per node.

The main interest of the Fermilab lattice group in using CANOPY and the ACPMAPS system is the application of lattice gauge theory to QCD and "beyond the standard model" phenomenology.[10] However, since site oriented problems involving numerical solutions of differential equations pervade all of science and engineering, the hardware and software we have just described clearly have a much broader applicability than just high energy theory.

4. FUTURE DIRECTIONS

To a large extent future ACP directions, given the apparently insatiable demand of high energy physicists for computer cycles, will be driven by the extremely fast pace at which microprocessor performance is increasing. RISC processors running at clocks upwards of 50 MHz, with performance at 50 and even 100 MIPS now seem assured within only 2 years or so. Technically this puts a big demand on cache systems and on the speed of static RAMS used in caches. This problem must be solved by the processor companies since it is universal for their customers.

The other issue the extraordinary performance of these devices will raise in multiprocessor environments is the inadequacy of busses to support their communication requirements. It will not be a matter of picking a slightly faster bus than VME (like Multibus II or even Fastbus). The bus concept itself will be obsolete and will have to be replaced by point to point communication. For the ACP this appears to mean that future processors, even those for experimenters, will have to plug directly into the Branchbus Switch Crate. This will provide the communication bandwidth they need, but no longer in a commercial crate standard where the marketplace delivers a large and always improving variety of I/O controllers and other essential peripheral devices. The Branchbus Switch is the same height as VME but deeper. It is therefore likely that the ACP will develop an interface card that will allow VME devices to be plugged transparently into the Branchbus Switch.

With very high performance RISC processors in the Branchbu Switch Crate, the experimenter's ACP system seems likely to merge with that designed for theorists. However, the processors may continue to differ because large theoretical problems, with their very regular accesses to memory, tend to have a very high cache miss rate in conventional cache designs. Multi level cache may be the answer to this problem in an approach that could be equally effective for both classes of HEP problems. (In fact this may be the way in which manufacturers solve the static RAM speed requirements mentioned earlier in an affordable way.) An alternative approach, which adds significant software and hardware complexity, could be called "anticipatory" cache. In regular theoretical problems the data that will be required from memory is known well ahead of time. Means could be devised to have programs inform the hardware of anticipated memory fetches so the data could be moved from slow memory to cache before it is needed.

The way in which the huge anticipated increases in computing are going to be brought into on line systems for data acquisition and triggering for the high rate SSC era detectors is very likely going to be the subject of ACP development work over the next few years. There is also underway a project to develop particularly efficient work station tools for doing analysis. Clearly when the new processor systems reduce the turn around time to pass through a DST data base from days to less than an hour, it will be inappropriate to continue spending days lining up calls to HBOOK (a histogramming package) for a next analysis run. Macintosh like human interfaces, adjusted to physicist needs and abilities, will be used on Apple or, perhaps, Sun or other workstations.

It is clear that the opportunity exists to continue development of usable extremely high performance, yet affordable, parallel machines for high energy physics and other sciences hungry for computer power. This opportunity is really an obligation given the strongly felt need.

5. REFERENCES

a. Permanent address: Centro Brasileiro de Pesquisas Físicas; Rua Xavier Sigaud 150; Rio de Janeiro, RJ, 22290 Brazil.

b. Permanent address: University of Arizona, Tucson, AZ 85721

1. Gaines, Irwin and Nash, Thomas, Use of New Computer Technologies in Elementary Particle Physics, *Ann. Rev. Nucl. Part. Sci.* 37, 177-212 (1987)

2. Fox, G.C., Otto, S.W. *Phys. Today* 37(5): 50-59 (1984); Fox,G. The performance of the Caltech Hypercube in scientific calculations. In *Supercomputers - Algorithms, Architectures and Scientific Computation*, ed. F.A. Masten, T. Tajima, Univ. Texas (1985); Seitz, C.L., *Commun. ACM* 28 22 (1985).

3. Kunz, P. F., *Nucl. Instrum. Methods* 135 435-440 (1976); Kunz, P.F., Gravina, M., Oxoby, G., Trang, Q., Fucci, A., Jacobs, D., Martin, B., and Storr, K., The 3081/E Processor in *Proc. Three Day In-Depth Rev. on the Impact of Specialized Processors in Elementary Part. Phys.*, Padova, 1983, 83-100 (1983).

4. Gaines, I., Areti, H., Atac., R., Biel, J., Cook, A., Fischler, M., Hance, R., Husby, D., Nash, T., and Zmuda, T., The ACP Multiprocessor System at Fermilab, *Comp. Phys. Comm.* 45 323-329 (1987); Biel, J., Areti, H., Atac., R., Cook, A., Fischler, M., Gaines, I.,Hance, R., Husby, D., Nash, T., and Zmuda, T., Software for the ACP Multiprocessor System,*Comp. Phys. Comm.* 45 331-3379 (1987).

5. Hance, R., Areti, H., Atac, R., Biel, J., Cook, A., Fischler, M., Gaines, I., Husby, D., Nash., T., and Zmuda, T., The ACP Branchbus and Real Time Applications of the ACP Multiprocessor System, *IEEE Trans. Nucl. Sci,* NS-34, 878-883 (1987).

6. *Second Generation ACP Multiprocessor System: System Specification Document*; July 25, 1988 and revisions; Fermilab Advanced Computer Program internal document; unpublished. Availability of ACP publications may be determined by reference to the file at the HEPNET location, `fnacp::acpdocs_root:[docs]doclist.doc`

7. Gottlieb, S.A. , et al., *Phys. Rev. Lett.* 55 1958 (1985); Christ, N.H. and Terrano, A.E., *Phys. Rev. Lett.* 56 111 (1986).

8. Christ, N.H. and Terrano, A. E., *IEEE Trans. Comp.* C-33(4) 344 (1984); Beteem, J., Denneau, M. and Weingarten, D., *J. Stat. Phys.*, 43 1171 (1986); Albanese, M., et al., The APE Computer, ROM2F/87/005 (1987).

9. *CANOPY -- ACP MAPS Software - Specifications Document*, Fermilab Advanced Computer Program internal document; March 26, 1988 and revisions; unpublished.

10. Mackenzie, P., Eichten, E., Hockney, G., Thacker, H.B., Atac, R., Cook, A., Fischler, M., Gaines, I., Husby, D., Nash, T., ACPMAPS: The Fermilab Lattice Supercomputer Project, Proceedings International Symposium *Field Theory on the Lattice*, Seillac, France, Sept. 28- Oct. 2, 1987, Nucl. Phys.B (Proc. Suppl.)4, pp 580-584 (1988).

COMPUTERS AND MOLECULAR QUANTUM MECHANICS: 1965-1988, A PERSONAL PERSPECTIVE

HENRY F. SCHAEFER III
Center for Computational Quantum Chemistry*, School of Chemical Sciences, University of Georgia, Athens, Georgia, U.S.A.

Abstract From the earliest days of electronic computers, molecular electronic structure theorists have been among the most demanding of users, both in terms of raw cpu power and machine sophistication. This report reviews computer hardware and software developments during what may be called the Second and Third Ages of Quantum Chemistry.

INTRODUCTION

I would like to thank Dr. Chris Bottcher for the opportunity to address this gathering of distinguished theoretical atomic and nuclear physicists. I actually began my scientific life as an atomic physicist, publishing 10 of my first 17 papers[1] in the Physical Review and Physical Review Letters. Both my undergraduate degree from M.I.T. and my graduate degree from Stanford were granted in the interdisciplinary area of Chemical Physics. Leonard Schiff was the most distinguished member of my Ph.D. thesis examination committee.

In fact, I was hired at Berkeley 19 years ago as an assistant professor because of the support of the nuclear chemists (who we know are really nuclear physicists) at the Lawrence Radiation Laboratory (LRL). Those nuclear physicists were persuaded by our research

*Contribution CCQC No. 23

at the time on atomic hyperfine structure[2] and the determination of nuclear moments[3].

I must confess that the siren call of chemistry began to strongly attract me[4] in 1969, and I published what I thought was my last atomic physics paper[5] in 1971. However last year we did complete some research[6] on the hyperfine structure of the (2p3p) $^1P_1$ state of N^+, leading to an improved value for the ^{14}N nuclear quadrupole moment.

COMPUTERS OF MY YOUTH

My experience in computing was relatively uneventful until 1972. As an M.I.T. undergraduate, I began in 1965 on the IBM 7094, which for its time was a supercomputer. Interestingly enough, M.I.T. has never institutionally supported large scale scientific computations since the departure 20 years ago of the IBM 7094. At M.I.T. I learned both FORTRAN and the IBM machine language FAP, and applied these to my senior thesis[7], on the molecular electronic structure of the H_8 cube.

Moving to Stanford for graduate study in 1966, I was very comfortable with the IBM 7090 entrenched at the computer center. However, my experience with FAP became history when a new IBM 360/67 machine was installed at Stanford in 1967. In fact, most of my thesis research was done on the 360/67 at Stanford. However, one particularly challenging computation[8] (on the ground state of the neon atom) was carried out on the IBM 360/195, a genuine supercomputer at the time (1969). That computation was made possible by the cooperation of Bob Nesbet, who remains a distinguished senior scientist at the IBM Research Laboratory in San Jose.

While still a graduate student at Stanford, a fruitful collaboration developed with Steve Rothenberg, a quantum chemist with University Computing Company (UCC) in Palo Alto. In 1970 Dr. Rothenberg hired Charlie Bender and the three of us made abundant use of the UCC Univac 1108 computer. In fact all the computations for what became my most influential paper[9] were carried out on the Univac 1108.

One of the strong attractions of the professorship at Berkeley was the committment by Isadore Perlman (then head of the Nuclear Chemistry Division of the Lawrence Radiation Laboratory) of 100 hours per year of time on the Control Data Corporation (CDC) 6600 machine. The 6600 was replaced at LRL in 1971 by the CDC 7600, the state-of-the-art supercomputer of that era. I had already acquired some experience on the 7600 at the Lawrence Radiation Laboratory in Livermore, where a rich collaboration existed with my friend Charlie Bender, now director of the State of Ohio Supercomputer Center.

The CDC 7600 served the needs of my research group superbly well until July of 1972. At that time the Lawrence Radiation Laboratory decided that the computer center would have to "pay its own way", i.e., recover sufficient revenues to pay the expenses of operating the CDC 7600. This decision resulted in our computing budget being exhausted during the first two months of the fiscal year. Although I pleaded "hardship" all the way up to Glenn Seaborg (recently returned to Berkeley from Washington as head of the old Atomic Energy Commission), no relief was to be found.

Parenthetically one of the factors leading to the demise (in 1980) of the visionary National Resource for

Computation in Chemistry (NRCC) was an analogous stinginess by LRL in providing CDC 7600 time to a young division that should have been nurtured and protected. I cannot think of a single circumstance in which the "pay hard cash as you go" philosophy of computing succeeded in university life, either financially or scientifically.

THE MINICOMPUTER ADVENTURE

While we intensified our collaborative efforts with Charlie Bender at Livermore and Steve Rothenberg at University Computing, I was forced to search for new computing resources. During a one-week visit to Harvard (all assistant professors should make such a pilgrimmage)in Autumn, 1972, a possible solution emerged. The Harvard Chemistry Department was in the process of purchasing a Digital Equipment Corporation (DEC) PDP-11/45. The principal purpose for the new Harvard minicomputer was to carry out the trivial arithmetic computations required by graduate students in their coursework. However Dick Stevens and Roy Gordon let me know that they hoped to try to carry out some substantive scientific computations on their new minicomputer. Incidentally, the latter efforts on the PDP 11/45 never bore fruit.

Returning with much enthusiasm to Berkeley, I told my colleague Bill Miller of my minicomputer discovery. He was in equally severe straits as far as computing resources were concerned (his postdocs Tom George and Jim Doll had ravenous cpu appetites) and encouraged me to try to put together a proposal. Our first version of the proposal was for two DEC PDP 11/45 computers and was greeted with great hostility by the Berkeley campus computing committee (which included Sid Fernbach, then head of all computing at Livermore). In fact, they refused to allow us to submit the proposal to the National Science Foundation (NSF). Fortunately Glenn Seaborg, who was personally aware of the gravity of our computing situation, had their decision overturned.

While we were engaged in battle with the Berkeley campus computer committee, it came to my attention that a small company named Datacraft was attempting to deliver a minicomputer that was significantly faster and more cost effective than the DEC PDP 11/45. Before the proposal[10] actually left the Berkeley campus for the NSF in Washington, the critical switch from DEC to Datacraft was accomplished.

Our NSF proposal "Large Scale Scientific Computation Via Minicomputer" was much more radical at the time than most persons can imagine now. Although the Datacraft 6024/4 was about the same price and performance as the DEC VAX 11/780, it must be remembered that our proposal was submitted four years before the introduction by DEC of the VAX. My friend Steve Rothenberg (one of the very few chemists at the time involved in commercial computing) stated confidently "It will never work, but at least your graduate students will require some marketable skills".

By all accounts [Computer World, March 8, 1976; A. L. Robinson, Science 193, 470 (1976); W. G. Richards Nature 266, 18 (1977).], the Miller/Schaefer minicomputer experiment was a remarkable success. It led directly to the purchase of subsequent generations of minicomputers by many research groups and chemistry and physics departments in universities around the world. I was even asked to present a plenary lecture[11] in September of 1975 to as skeptical a group as the IEEE Computer Society at their annual meeting in Washington, D.C.

BERKELEY TO TEXAS TO BERKELEY

In September, 1979 I became the Wilfred T. Doherty Professor of Chemistry (following Dick Bernstein's departure to Columbia) and inaugural Director of the Institute for Theoretical Chemistry at the University of Texas, Austin. One of the considerations that had some effect on the decision to leave Berkeley was the age (six years!) of the Datacraft 6024/4 minicomputer. At Texas, our research group's arrival coincided within one week of the installation of a new CDC Cyber 170/750 mainframe. The promise of 25 hours per week on the Cyber dwarfed our capacity on the Datacraft minicomputer at Berkeley.

One consideration that made possible my group's return to Berkeley one year later was the university's purchase of a new Harris Corporation 800 minicomputer, dedicated to the exclusive use of the research group's of Bill Miller and myself. The Harris 800 proved a good machine, serving us well until late 1983. One significant computational achievement on the Harris 800 was the first configuration interaction (CI) wave function involving more than <u>one million</u> parameters determined with full variational rigor.[12]

The shorter lifetime of the Harris 800 (less than four years) compared to the Datacraft machine (six years) is symptomatic of an accelerating trend. It may be argued that the useful lifetime of a state-of-the-art minicomputer is now only two years. Bill Miller and I submitted a new minicomputer proposal to the NSF in August of 1983 and it was funded in early 1984. The new machine, the Gould 32/8780 proved extremely disappointing during the eight months it resided in the chemistry department at Berkeley. The experience also brought to my attention the much greater need for user-friendliness with the enormous expansion of our working quantum chemistry codes during the period 1973-83.

By the middle of 1984 we had effectively wasted one-half year of research time on the Gould and were forced by circumstances to go with a much more conservative machine, either DEC or IBM. We also extensively benchmarked the ELXSI machine, but rejected it as being potentially too risky. Prior to 1983 I had the opinion (now seen as arrogance) that we could harness any computer (no matter how unruly) to the quantum chemistry plow. The Gould disaster was very sobering.

THE MOVE TO IBM

The winner in the DEC/IBM competition was the IBM 4381 Model Group 2, which arrived in October, 1984. The 4381 was an order of magnitude more user-friendly than any machine our group had experienced and was immediately popular with the Miller/Schaefer graduate students and postdoctorals (about 25 in all).

At the same time, great changes were taking place in computing on the Berkeley campus. Ray Neff arrived as campus-wide computer czar, and IBM gave Berkeley a well-equipped dual processor 3081 mainframe. Dr. Neff encouraged us to use the excess cycles on the IBM 3081 and in short order my research group was getting 100 hours per week regularly. The euphoria increased when Chancellor Michael Heyman announced in 1986 that Berkeley would purchase a CRAY XMP supercomputer in addition to the IBM 3081. I was not personally interested in the Berkeley CRAY XMP because our codes were by this time quite IBM-dependent, and the Berkeley CRAY was to use the untested UNICOS operating system.

Roughly coincident with the CRAY announcement, Ray Neff informed me that he also intended to upgrade the IBM 3081 to the new IBM 3090. One must remember that the Berkeley quantum chemistry group was thoroughly addicted to the 3081 by this time, in fact doing twice as much chemistry on the 3081 as on our own IBM 4381-MG2. At this point Neff brought to bear a carrot-and-stick strategy. The IBM 3081 was going to disappear irrespective of other future decisions. Neff wanted to sell our IBM 4381 (which was still worth about $400,000) and use the cash as an initial payment on the IBM 3090-200.

Our options were (a) sell the 4381 and become first partner in the new 3090, effectively doubling our computing resources; (b) keep the 4381, losing access to the departing 3081, and thus reducing our computing resources by two-thirds. Although we were very uncomfortable about giving up the autonomy of our own IBM 4381-MG2, we had little practical option to Neff's proposal for 3090 partnership. Fortunately, the 3090 partnership proved a very fruitful one, with my research group averaging more than 160 hours of cpu time per week for the period October, 1986 - August 1987.

THE CENTER FOR COMPUTATIONAL QUANTUM CHEMISTRY

In March of 1987 it was announced that Dr. Charles B. Knapp would become the new president of the University of Georgia on July 1. My friend Charlie Bender (then

Director of the Advanced Computational Methods Center) had earlier attempted to recruit me to Georgia. However the absence of IBM 3090 power and of a permanent president at Georgia made the offer unsuccessful. President-elect Knapp assured me in April 1987 that the purchase of an IBM 3090-200 E (about 15% faster than the Berkeley machine) was imminent and committed Georgia to the establishment of a new Center for Computational Quantum Chemistry (CCQC).

In August of 1987 I became Graham Perdue Professor of Chemistry and Director of the CCQC at the University of Georgia (UGA). One month later the Georgia IBM 3090-200E (delivered in July) was upgraded to a four processor IBM 3090-400E. Despite the massive 3090 cpu power at Georgia, the first six months there were painful for the Berkeley-transplanted group, due to a severe shortage of disk space and disk controllers. However, these problems were resolved in March, 1988, and the CCQC is now receiving IBM machine power comparable to that appreciated in Berkeley.

Early this month (April, 1988) IBM announced a "jointly defined effort" with the Center for Computational Quantum Chemistry. The CCQC will optimize its codes for the IBM 3090 vector facility and make them available (in executable form) to IBM users. Prior to this, the Schaefer group codes have only been made available to former members of the group. Carefully optimized quantum chemistry code for the 3090 vector facility runs at speeds comparable to vectorized CRAY-XMP code.[13] In return IBM provided more than one million dollars worth of hardware, software, and maintainence to CCQC.

The centerpiece of the new IBM equipment is the 9370 Model 90, a machine roughly 10% of the speed of the 3090 but supporting all of the 3090 software that is integral to our current codes. The IBM 9370-90 will be located in the CCQC and used for code development and file editing, prior to sending large jobs to the UGA campus 3090-400E.

EXPERIENCE WITH THE INTEL HYPERCUBE

Before proceeding to organic chemistry, some discussion of hypercubes is in order, given the orientation of the present conference. Our Berkeley group (in collaboration with Sandia Livermore) was the first to use the Intel Hypercube for quantum chemistry,[14] the work resulting in Mike Colvin's Ph.D. thesis. To our knowledge no other quantum chemistry groups have successfully used the hypercube to date.

To explain the absence of an exodus of quantum chemists to the hypercube, one should say something about the nature of quantum chemistry computer programs. The Berkeley/Georgia IBM programs consist of about 900,000 active lines of FORTRAN code, plus some critical assembly language input/output routines. Of this code about 200,000 lines are used on a sustained, regular basis. The codes include methods for self-consistent-field (SCF) total energies and analytic first[15], second[16], and third[17] energy derivatives; multiconfiguration SCF energies and derivatives[18]; configuration interaction (CI) total energies[12,19] and analytic first[20] and second[21] energy derivatives; coupled cluster total energies[22] and analytic energy first derivatives[23]; and a variety of molecular properties including infrared[24] and Raman[25] vibrational spectral intensities.

For quantum chemists to use the Intel hypercube effectively most of the algorithms must be completely redesigned. It took Mike Colvin, a highly talented and computer-wise graduate student, about one year to develop a general self-consistent-field (SCF) total energy program. Unfortunately, the SCF total energy represents less than 5% of the CCQC active codes. Even so, such an enormous investment in hypercube algorithms might be acceptable if quantum chemistry methodology was stationary.

In fact, nearly our entire library of quantum chemistry codes turns over about every five years. Staying at the forefront of the field thus requires an enormous theoretical/methodological/computational effort, even for a group such as CCQC with 15 active scientists. Until quantum chemistry software reaches a stationary point (perhaps the year 2000?) the sort of complete algorithm redesign necessary for the Intel hypercube cannot be a priority for the bulk of our computer codes. Should some sort of super-smart FORTRAN compiler be developed for hypercubes, the above problems would of course disappear. However, we see no such compiler on the immediate horizon.

1988 SUPERCOMPUTER COMMENTS

(1) For the foreseeable future, the latest, most cost-effective minicomputer or microcomputer will outrun the most cost effective supercomputer. As an example we may compare the SUN 4 machine to the CRAY-XMP. The Dongarra LINPACK algorithm,[26] highly favorable to the CRAY in this comparison runs about 50 times faster on the CRAY-XMP than on the SUN 4. However, the one-processor XMP is more than 50 times more expensive than the SUN 4. Moreover, for many quantum chemistry codes the CRAY-XMP is much less than 50 times faster than the SUN 4. Most

notably, for SCF total energy computations, the CRAY-XMP is less than 10 times faster than the SUN 4.

It appears likely that the above trend will continue into the foreseeable future. The just-announced[27] "desktop supercomputers" from Apollo, Ardent, and Stellar are significantly more cost-effective than the SUN 4. In contrast, progress from the CRAY-XMP to the not-yet-delivered CRAY-YMP has been painfully slow.

(2) For the past three years, political considerations in the U. S. have caused supercomputers to be very attractive.

(a) The establishment of NSF supercomputer centers have provided reasonable amounts of "free" time to qualified chemists and physicists. Unfortunately the NSF, under great budgetary pressures, does not appear prepared to move ahead with necessary second-round equipment upgrades for the five national centers. The impending move of supercomputer-advocate Kenneth Wilson from the NSF-supported Cornell supercomputer center to the State of Ohio Supercomputer Center at Columbus would appear to indicate which way the wind is blowing.

(b) The U. S. government has an obvious and justifiable stake in keeping the CDC/ETA supercomputer venture alive. The purchase of ETA-10 machines for Florida State and Princeton would not otherwise be justifiable in light of the balky software performance of the CDC Cyber 205 predecessors. However, the much lower price and maintenance costs for the new air-cooled ETA-10 units may breathe new life into CDC/ETA.

(c) University IBM 3090/Vector Facility installations benefit enormously (in terms of overall cost effectiveness) from their ability to service a vast number of daytime interactive users with consistent, high-level software. After all, the 3090 was really designed for banks. The above university bread-and-butter interactive usage may be readily combined with that of a few groups (such as CCQC at Georgia) whose research justifies the use of large amounts of off-peak capacity.

(d) The natural, justifiable, predominant scientific usage of supercomputers in the foreseeable future is at least 70% molecular modeling, with chemistry occupying the central position. Chemistry flows in one direction toward biochemistry and molecular biology and in the other toward applied physics and materials science. Electronic structure is the common thread connecting all these areas. In fact, molecular modeling currently offers significantly more job opportunities than quantum chemistry is producing Ph.D.'s.

AN EXAMPLE: THE [n] PARACYCLOPHANES

These molecules consist of a benzene ring linked at the opposite carbon atoms by a chain of methylene (CH_2) groups. Thus, for the molecule [n] paracyclophane

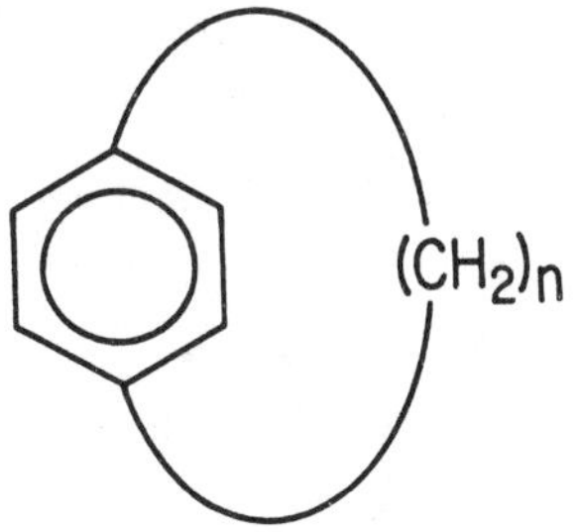

the connecting methylene chain is $(CH_2)_n$. These compounds have been of interest to organic chemistry for 40 years, with 1987 Nobel Prize winner Donald Cram playing a pioneering role[28] in the 1950's. A fundamental question in organic chemistry is whether these bent-benzene molecules are "aromatic".

With time provided by the San Diego Supercomputer Center Allocation Committee, we have carried out ab initio quantum mechanical studies of the molecules [5], [6], [7], [8], and [9] paracyclophane. The technical details of these theoretical studies are summarized in the Table I.

TABLE I Technical details of ab initio quantum mechanical studies of the [n] paracyclophane molecules. Note that both minimum (STO-3G) and double zeta [c(9s5p/4s2p), H(4s/2s)] basis sets have been used for n = 5,6,7.

[n]	Molecule	Basis Functions	Geometry Optimized	Force Constants Evaluated
5	$C_{11}H_{14}$	69	x	x
		138	x	x
6	$C_{12}H_{16}$	76	x	x
		152	x	x
7	$C_{13}H_{18}$	83	x	x
		166	x	No
8	$C_{14}H_{20}$	90	x	x
9	$C_{15}H_{22}$	97	x	No

The research on the [n] paracyclophanes represents state-of-the-art ab initio molecular quantum mechanics. However, this requires some explanation, since the most complicated wave function considered involves "only" 166 contracted gaussian basis functions. In comparison, the brilliant efforts of Jan Almlöf at the University of Minnesota Supercomputer Center have treated molecules with more than 1000 basis functions.[29]

It may first be noted that Almlöf's large molecules have very high point group symmetries. In contrast the

[n] paracyclophanes have only one element of symmetry, either a plane of symmetry or a two-fold axis of rotation. Secondly the Almlöf studies involve at most computation of a few total energies only. The present work on the [n] paracyclophanes requires total energies, but also first and second energy derivatives with respect to all nuclear coordinates.

In addition, for each of the paracyclophanes studied, a complete optimization of all geometrical parameters was carried out. The equilibrium geometry for [7] paracyclophane is illustrated in Figures 1 and 2.

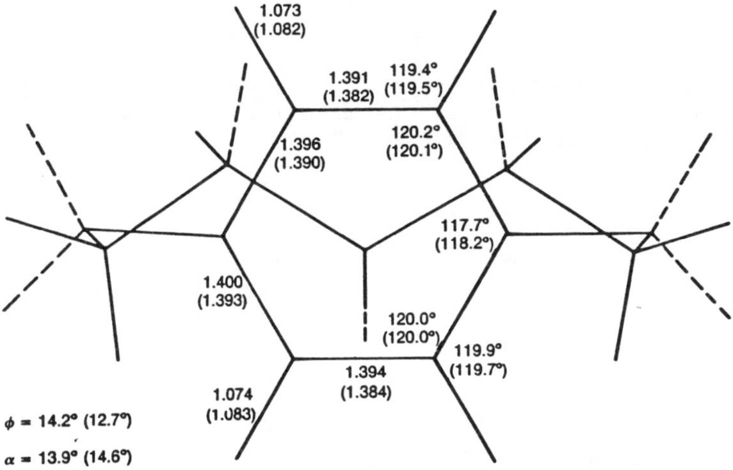

FIGURE 1. Theoretical molecular structure of [7]paracyclophane based on the double zeta (DZ) self-consistent field (SCF) method. This sketch shows only the geometrical parameters for the benzene ring portion of the molecule. In parentheses the minimum basis set (MBS) SCF structural parameters are given.

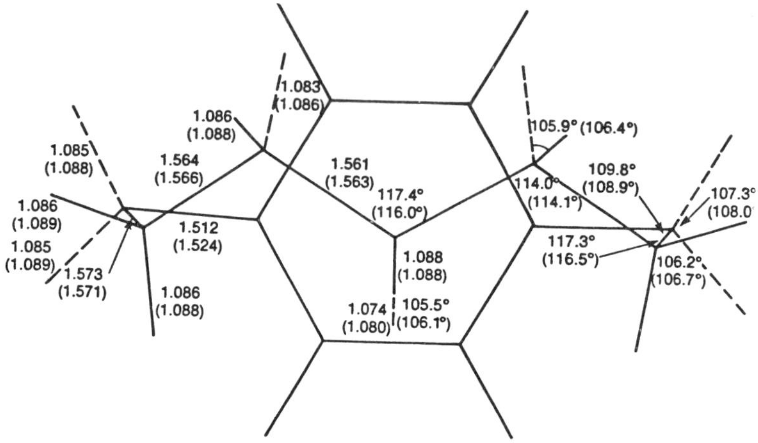

FIGURE 2. Other features of the DZ SCF theoretical molecular structure for [7]paracyclophane. Shown here are geometrical parameters for the seven methylene groups [i.e., $(CH_2)_7$] bridging the para positions of the benzene ring. In parentheses the MBS SCF values are given.

Low-symmetry force constant evaluations with 152 basis functions and geometry optimizations with 166 basis functions are close to the limit achievable with 1988 computing machinery. Comparable studies of different classes of molecules have been carried out by Stephen Binkley at Sandia Livermore and by David Dixon at DuPont. Both workers used CRAY supercomputers located at their own research cites.

The [n] paracyclophanes were studied using a 135 hour allocation of XMP time. In fact, by using low priorities (during the early days at the San Diego Supercomputer Center), this 135 hour CRAY-XMP allocation was leveraged into about 500 actual CPU hours. The load at SDSC during the past year has made it very difficult to compute at such low priorities. In addition, the enormous current demand for disk space at SDSC might make it physically impossible to run the larger [n] paracyclophane computations there in 1988.

SUMMARY

Quantum chemists have been forced to be exceptionally resourceful in their search for cpu cycles over the past 20 years. Our experience is typical in some respects and unique in others. We hope to continue for many years the computational quest for more and more accurate solutions of Schrödinger's Equation for ever larger molecular systems.

ACKNOWLEDGEMENTS

We are sincerely appreciative of the continuous support of the National Science Foundation and Department of Energy for many years.

REFERENCES

1. See for example, H. F. Schaefer and F. E. Harris, "Metastability of the ^1D State of the Nitrogen Negative Ion", Phys. Rev. Letters, 21, 1561 (1968).
2. H. F. Schaefer, R. A. Klemm, and F. E. Harris, "Atomic Hyperfine Structure I. Polarization Wave Functions for the Ground States of B,C,N,O, and F", Phys. Rev., 176, 49 (1968).
3. H. F. Schaefer and R. A. Klemm, "Atomic Hyperfine Structure III. Excited States of C,N, and O", Phys. Rev., 188, 152 (1969).
4. S. Rothenberg, R. H. Young, and H. F. Schaefer, "Ground State Self-Consistent-Field Wave Functions and Molecular Properties for the Isoelectronic Series SiH_4, PH_3, H_2S, and HCl", J. Amer. Chem. Soc., 92, 3243 (1970).
5. O. R. Platas and H. F. Schaefer, "Symmetry-Adapted Pair Correlations for Open-Shell Systems: Correlation Energy and Hyperfine Structure of the Nitrogen Atom", Phys. Rev.A. 4, 33 (1971).

6. G. E. Scuseria and H. F. Schaefer, "The Nuclear Quadrupole Moment of ^{14}N. A Theoretical Prediction from Full Valence Shell and Full Configuration Interaction Atomic Wavefunctions", J. Chem. Phys., 87, 4020 (1987).
7. H. F. Schaefer, "A Study of the H_8 Cube Using the Molecular Orbital and Extended Valence Bond Theories", S. B. Thesis, Massachusetts Institute of Technology, June, 1966.
8. J. W. Viers, F. E. Harris, and H. F. Schaefer, "Pair Correlations and the Electronic Structure of Neon", Phys. Rev. A, 1, 24 (1970).
9. C. F. Bender and H. F. Schaefer, "New Theoretical Evidence for the Nonlinearity of the Triplet Ground State of Methylene", J. Amer. Chem. Soc., 92, 4984 (1970).
10. That original proposal and a subsequent assessment of the Datacraft experiment is publicly available as Report PB-267-575, National Technical Information Service, Department of Commerce, Springfield, Virginia.
11. H. F. Schaefer, "Are Minicomputers Suitable for Large Scale Scientific Computations?" Proceedings of the Eleventh Annual IEEE Computer Science Conference (Washington, D.C., September, 1975), pages 61-64.
12. P. Saxe, D. J. Fox, H. F. Schaefer, and N. C. Handy, "The Shape-Driven Graphical Unitary Group Approach to the Electron Correlation Problem. Application to the Ethylene Molecule", J. Chem. Phys., 77, 5584 (1982).
13. B. Liu, Comparison of GAUSSIAN 86 on IBM 3090/Vector and CRAY-XMP, personal communication, August, 1987.
14. M. E. Colvin, "Quantum Chemical Methods for Massively Parallel Computers", Ph.D. Thesis, University of California, Berkeley, November, 1986; a preliminary report is given by R. A. Whiteside, J. S. Binkley, M. E. Colvin, and H. F. Schaefer, "Parallel Algorithms for Quantum Chemistry. Integral Transformations on a Hypercube Processor", J. Chem. Phys., 86, 2185 (1987).
15. J. D. Goddard, N. C. Handy and H. F. Schaefer, "Gradient Techniques for Open-Shell Restricted Hartree-Fock and Multi-Configuration Self-Consistent-Field Methods", J. Chem. Phys. 71, 1525 (1979).
16. (a) P. Saxe, Y. Yamaguchi and H. F. Schaefer, "Analytic Second Derivatives in Restricted Hartree-Fock Theory. A Method for High-Spin Open-Shell Molecular Wave Functions", J. Chem. Phys. 77, 5647 (1982);

(b) Y. Osamura, Y. Yamaguchi, P. Saxe, M. A. Vincent, J. F. Gaw and H. F. Schaefer, "Unified Theoretical Treatment of Analytic First and Second Energy Derivatives in Open-Shell Hartree-Fock Theory", Chem. Phys. 72, 131 (1982).

17. (a) J. F. Gaw, Y. Yamaguchi, and H. F. Schaefer, "Analytic Third Derivatives for Self-Consistent-Field Wavefunctions", J. Chem. Phys. 81, 6395 (1984);
(b) J. F. Gaw, Y. Yamaguchi, H. F. Schaefer and N. C. Handy, "Generalization of Analytic Energy Third Derivatives for the RHF Closed-Shell Wavefunction: Derivative Energy and Integral Formalisms and the Prediction of Vibration-Rotation Interaction Constants", J. Chem. Phys. 85, 5132 (1986);
(c) J. F. Gaw, Y. Yamaguchi, R. B. Remington, Y. Osamura, and H. F. Schaefer, "Analytic Energy Third Derivatives for Open-Shell Self-Consistent-Field Wavefunctions", Chem. Phys. 109, 237 (1986).

18. (a) Y. Yamaguchi, Y. Osamura, G. Fitzgerald and H. F. Schaefer, "Analytic Force Constants for Post-Hartree-Fock Wave Functions: The Simplest Case", J. Chem. Phys. 78, 1607 (1983);
(b) M. R. Hoffmann, D. J. Fox, J. F. Gaw, Y. Osamura, Y. Yamaguchi, R. S. Grev, G. Fitzgerald, H. F. Schaefer, P. J. Knowles and N. C. Handy, "Analytic Energy Second Derivatives for General MCSCF Wavefunctions", J. Chem. Phys. 80, 2660 (1984);
(c) M. R. Hoffmann and H. F. Schaefer, "MCSCF Energy Derivatives Using Fock Operator Methods", pages 63-77 of "Geometrical Derivatives of Energy Surfaces and Molecular Properties", editors P. Jorgensen and J. Simons (D. Reidel, Dordrecht, Holland, 1986);
(d) M. Duran, Y. Yamaguchi, R. B. Remington, and H. F. Schaefer, "Analytic Energy Second Derivatives for Paired-Excited Multi-Configuration Self-Consistent-Field Wavefunctions. Application of the PEMCSCF Model to H_2O, CH_2, HCN, HCCH, H_2CO, NH_3, CH_4, and C_2H_4", Chem. Phys., in press.

19. (a) B. R. Brooks and H. F. Schaefer, "The Graphical Unitary Group Approach to the Electron Correlation Problem. Methods and Preliminary Applications", J. Chem. Phys. 70, 5092 (1979);
(b) B. R. Brooks, W. D. Laidig, P. Saxe, N. C. Handy and H. F. Schaefer, "The Loop-Driven Graphical Unitary Group Approach: A Powerful Method for the Variational Description of Electron Correlation", Phys. Scr. 21, 312 (1980)

20. (a) B. R. Brooks, W. D. Laidig, P. Saxe, J. D. Goddard, Y. Yamaguchi and H. F. Schaefer, "Analytic Gradients from Correlated Wave Functions via the Two-Particle Density Matrix and the Unitary Group Approach", J. Chem. Phys. 72, 4652 (1980);
(b) J. E. Rice, R. D. Amos, N. C. Handy, T. J. Lee, and H. F. Schaefer, "The Analytic Configuration Interaction Gradient Method: Application to the Cyclic and Open Isomers of the S_3 Molecule", J. Chem. Phys. 85, 963 (1986);
(c) T. J. Lee, W. D. Allen, and H. F. Schaefer, "The Analytic Evaluation of Energy First Derivatives for Two-Configuration Self-Consistent-Field Configuration Interaction (TCSCF-CI) Wave Functions. Application to Ozone and Ethylene", J. Chem. Phys. 87, 7062 (1987).
21. (a) D. J. Fox, Y. Osamura. M. R. Hoffmann, J. F. Gaw, G. Fitzgerald, Y. Yamaguchi and H. F. Schaefer, "Analytic Energy Second Derivatives for General Correlated Wavefunctions, Including a Solution of the First-Order Coupled-Perturbed Configuration Interaction Equations", Chem. Phys. Lett. 102, 17 (1983);
(b) T. J. Lee, N. C. Handy, J. E. Rice, A. C. Scheiner, and H. F. Schaefer, "The Efficient Evaluation of Configuration Interaction Analytic Energy Second Derivatives: Application to Hydrogen Thioperoxide, HSOH", J. Chem. Phys. 85, 3930 (1986).
22. G. E. Scuseria, A. C. Scheiner, T. J. Lee, J. E. Rice, and H. F. Schaefer, "The Closed-Shell Coupled Cluster Single and Double Excitation (CCSD) Model for the Description of Electron Correlation. A Comparison with Configuration Interaction (CISD) Results", J. Chem. Phys. 86, 2881 (1987).
23. (a) A. C. Scheiner, G. E. Scuseria, J. E. Rice, T. J. Lee, and H. F. Schaefer, "Analytic Evaluation of Energy Gradients for the Single and Double Excitation Coupled Cluster (CCSD) Wavefunction: Theory and Application", J. Chem. Phys. 87, 5361 (1987);
(b) G. E. Scuseria and H. F. Schaefer, "Analytic Evaluation of Energy Gradients for the Single, Double, and Linearized Triple Excitation Coupled Cluster CCSDT-1 Wavefunction: Theory and Applications", Chem. Phys. Lett. 146, 23 (1988).
24. Y. Yamaguchi, M. J. Frisch, J. F. Gaw, H. F. Schaefer, and J. S. Binkley, "Analytic Evaluation and Basis Set Dependence of Intensities of Infrared Spectra", J. Chem. Phys. 84, 2262 (1986).

25. M. J. Frisch, Y. Yamaguchi, J. F. Gaw and H. F. Schaefer, "Analytic Raman Intensities from Molecular Electronic Wave Functions", J. Chem. Phys. 84, 531 (1986).
26. J. J. Dongarra, Technical Memorandum No. 23, Mathematics and Computer Science Division Argonne National Laboratory, March 23, 1988.
27. See for example, front page story in the March 7, 1988 issue of Computerworld.
28. See, for example, D. J. Cram, N. L. Allinger, and H. Steinberg, "Macro Rings. VII. The Spectral Consequences of Bringing Two Benzene Rings Face to Face", J. Amer. Chem. Soc. 76, 6132 (1954).
29. H. P. Luthi and J. Almlöf, "Ab Initio Studies on the Thermodynamic Stability of the Icosahedral C_{60} Molecule Buckminsterfullerene", Chem. Phys. Lett., 135, 357 (1987).

AB INITIO CALCULATIONS OF PROCESSES IN SMALL MOLECULES

KATE P. KIRBY
Harvard-Smithsonian Center for Astrophysics,
60 Garden Street, Cambridge, Massachusetts, U.S.A.

Abstract Some basic molecular processes are reviewed with respect to the requirements for accurately quantifying them. Examples of photodissociaton, dissociative recombination and charge transfer are used to illustrate the difficulties and challenges for ab initio molecular structure methods and to highlight the areas in which the enhanced capabilities of supercomputers will therefore have considerable impact. Recent studies comparing the accuracy of various ab initio molecular structure methods with respect to correlation energy, and other properties are noted.

INTRODUCTION

A detailed understanding of a variety of molecular processes is fundamental to investigations of many macroscopic systems in biology, chemistry, aeronomy and astronomy. The speed and accessibility of supercomputer facilities affords the molecular physicist the opportunity to make advances in the study of molecular processes. In this paper I will describe a number of different molecular processes, and focus on three of them, giving examples of each. For the sake of simplicity and generality, I will use the canonical molecule, AB, to illustrate each process initially. There is the danger that all the processes may then seem deceptively simple. However, if one replaces every molecule I have written as AB with the molecule ABCDE... or larger, it is clear that the level of complexity increases very rapidly and that the capabilities of supercomputers are essential.

MOLECULAR PROCESSES

Photoprocesses

Absorption of a photon by a molecule AB may result in electronic, vibrational and rotational excitation:

$$AB + h\nu \rightarrow AB^* \quad .$$

If AB^* is an excited electronic state, unbound with respect to nuclear motion, as manifested by a repulsive potential energy curve, the molecule may fall apart in a process known as *photodissociation*:

$$AB^* \rightarrow A + B \quad .$$

If AB^* is an excited electronic state lying above the ionization threshold of AB, *autoionization* may occur after absorption of the photon:

$$AB^* \rightarrow AB^+ + e^-$$

Direct photoionization may also result from photon absorption:

$$AB + h\nu \rightarrow AB^+ + e^- \quad .$$

Electron-molecule Collision Processes

Collisions with electrons may excite molecule AB electronically, vibrationally or rotationally:

$$e^- + AB \rightarrow AB^* + e^- \quad .$$

Dissociative attachment may also occur, in which a negative ion is formed:

$$e^- + AB \rightarrow A^- + B \quad .$$

An electron may also attach to a positive molecular ion forming a dissociating resonance state, AB^{**}, which falls apart to form product atoms which may be electronically excited. This process is called *dissociative recombination*:

$$e^- + AB^+ \rightarrow AB^{**} \rightarrow A + B \quad .$$

Heavy-body Collision Processes

Collisions of atoms A and B can lead to molecule formation upon radiation of a photon through the *radiative association* process:

$$A + B \rightarrow AB^* \rightarrow AB + h\nu \quad .$$

Collisions between two atoms, one of which is electronically excited above the threshold for ionization of the other atom or the molecule AB, can give rise to either *associative ionization:*

$$A^* + B \rightarrow AB^{**} \rightarrow AB^+ + e$$

or *Penning ionization*:

$$A^* + B \rightarrow AB^{**} \rightarrow A + B^+ + e^- \quad .$$

The capture of an electron from a collision partner by an ion is known as *charge transfer*:

$$A^+ + B \rightarrow A + B^+ \; .$$

Collisions between an ion and a molecule can give rise to a variety of possible excited and rearranged products, for instance:

$$A^+ + BC \rightarrow AB^+ + C \quad .$$

Quantifying These Processes

In order to calculate accurately the cross sections, product distributions, or rates which characterize these molecular processes, one needs both the methods of molecular electronic structure and the methods of scattering theory. From ab initio molecular structure calculations one obtains accurate potential curves or surfaces, as well as electronic wavefunctions which then can be used to evaluate a variety of transition and interaction matrix elements of operators such as the dipole moment, the spin-orbit coupling, and the non-adiabatic Born-Oppenheimer coupling d/dR. Most commonly, such properties are then used in a scattering treatment appropriate to the particular molecular process. As suggested in the title of this paper, I will concentrate on the molecular structure aspects, primarily because, if the potential curves and interaction matrix elements are poorly determined, the cross sections will be worthless, no matter how accurately the scattering is treated.

I will focus briefly on three different processes: photodissociation, dissociative recombination, and charge transfer, highlighting the issues which make the electronic structure calculations difficult and challenging. Illustrative examples involving simple molecules will be used. As a preview, I list several recurring themes:

1) Potential curves over the full range of internuclear separations are usually needed, because of the scattering process; wavefunctions capable of describing accurately the molecular bonding region as well as the separated atom limit are essential.

2) There is a problem describing several states equally well. Traditional optimization methods to obtain molecular orbitals generally bias towards the lowest root of a symmetry. In order to make up for orbital deficiencies in excited states, many additional configurations may be necessary in the wavefunction.

3) Accurate descriptions of curve crossing regions are often essential because of extreme sensitivity of the cross sections to vibrational overlaps.

4) The accurate evaluation of complete coupling operators over a range of internuclear separations is required.

PHOTODISSOCIATION

An important process in certain regions of the earth's atmosphere and in diffuse interstellar clouds which may have enhanced radiation fields, photodissociation can occur through several different mechanisms which are illustrated in Figure 1. In each case photon absorption takes place in the lowest vibrational level (v=0) of the ground electronic state.

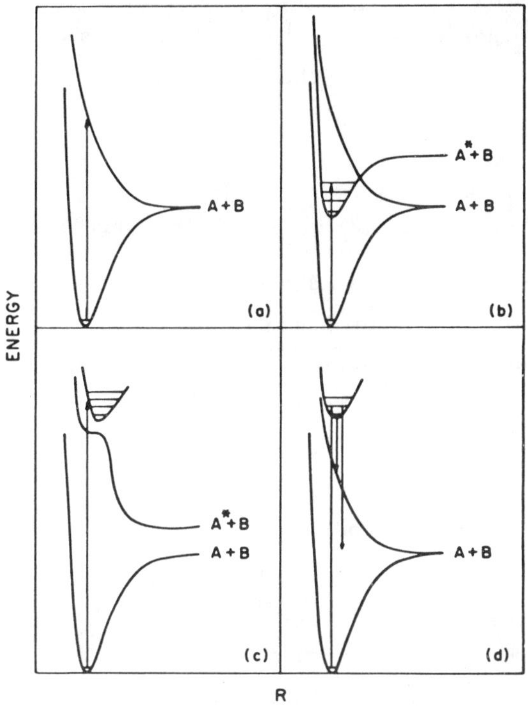

FIG 1. Various photodissociation processes of molecule AB; electronic potential energy as a function of internuclear separation. a) direct photodissociation; b) predissociation; c) coupled states photodissociation; d) spontaneous raditive dissociation. (from: van Dishoeck, reference 1).

Direct Photodissociation

In direct photodissociation an electronic transition is made into a repulsive excited state, (as shown in Figure 1a), or onto the repulsive wall of a bound excited state, above the dissociation limit. Within the framework of the Born-Oppenheimer approximation, the total wavefunction for a molecule can be written in terms of electronic ($\underline{r}$) and nuclear ($\underline{R}$) coordinates as $\psi(\underline{r},\underline{R}) = \psi_{el}(r|R)\zeta_{nuc}(\underline{R})$. The electronic wavefunction $\psi_{el}(\underline{r}|R)$ is a solution to the many-electron time-independent Schrödinger equation with eigenvalue $E_{el}(R)$ and $\zeta_{nuc}(R)$ is the wavefunction describing nuclear motion, the radial part of which, $\chi(R)$, is a solution to the one-dimensional radial Schrödinger equation:

$$-\frac{1}{2\mu}\frac{d^2}{dR^2}\chi(R) + [E_{el}(R) + \frac{1}{2\mu R^2}J(J+1)-E]\chi(R) = 0 \quad . \quad (1)$$

In the above μ is the reduced mass of the system and J is the angular momentum quantum number of the rotating molecule. The electronic energies, $E_{el}(R)$, form the potential within which the nuclei vibrate (if the state is bound) or dissociate (if the state is repulsive). The bound and continuum vibrational functions are denoted χ_v and χ_k, respectively. The direct photodissociation cross section, σ, is proportional to $\Delta E \cdot |\langle\chi_{v''}(R)|D_{fi}(R)|\chi_k(R)\rangle|^2$ where $D_{fi}(R)$ is the electronic dipole transition moment. The cross section is continuous as a function of photon energy.

The main inaccuracies in the calculation of cross sections for this process are in the determination of $E_{el}(R)$ and $D_{fi}(R)$. RKR curves can be used for the ground electronic state to help eliminate uncertainties, but these are rarely available for the upper electronic state in photodissociation.

An example of photodissociation to a number of higher-electronic states can be found in work by Kirby et al.[2] for CH^+ in which dissociation takes place through excited states which separate to $C(^1D)$ and $C(^1S)$ plus H^+. The ground state of CH^+ correlates to $C^+(^2P) + H$. Constructing a wavefunction to describe accurately all the various separated atom limits relevant to this problem is very difficult. However the correct description of the energy separations of the excited states is critical in evaluating the photodissociation rate in the interstellar medium for those channels lying near the Lyman cutoff of 912 Å.

Indirect Photodissociation

The remaining three photodissociation processes shown in Figure 1b), 1c), and 1d) are indirect processes which occur via discrete transitions to bound vibrational levels of an excited electronic state. The dissociation process then takes place upon radiation or through couplings to other states. For lack of space I will not elaborate on *spontaneous radiative dissociation* shown in Figure 1d, other than to note that its description is very similar to the direct photodissociation process, but with dissociation taking place on emission of a photon after the initial line absorption.

Molecular predissociation takes place when the excited electronic state is crossed by a dissociating state to which it is coupled through small terms in the molecular Hamiltonian which are often neglected. Terms such as spin-orbit, spin-spin and spin-rotation interactions as well as the kinetic energy of the rotating molecule, denoted as H_{int} in the following equation, are included within the framework of first-order perturbation theory. The predissociation rate of level i, v', J' by a final state, f, k, J is given by the Golden Rule formula[3]

$$k_{v'J'} = \frac{2\pi}{\hbar} |\langle \Psi_{iv'J'}(\underline{r}, \underline{R}) \, | H_{int} | \Psi_{fkJ}(\underline{r}, \underline{R}) \rangle |^2 \tag{2}$$

in which $\Psi_{iv'J'}$ (Ψ_{fkJ}) is the product of an electronic wavefunction and a bound (continuum) nuclear wavefunction. The difficulties associated with an accurate calculation of the rate are two-fold: first, accurate evaluation of small terms in the molecular Hamiltonian, such as spin-orbit, as a function of internuclear separation; and second, the extreme sensitivity of the above integral over nuclear coordinates to the relative positions of the interacting electronic states. This latter point means that the ab initio curves must cross correctly.

A good example of predissociation can be found in the work of van Dishoeck et al.[4] on the HCl molecule. The oscillator strength of the transition $C^1\Pi - X^1\Sigma^+$ is very large. The $C^1\Pi$ state is crossed near its R_e by the repulsive $^3\Sigma^+$ state, which is coupled to the $C^1\Pi$ state by the spin-orbit operator. Because of the large C-X oscillator strength, in a uniform radiation field the predissociation route turns out to be more efficient in destroying the molecule than the direct photodissociation channel through the low-lying repulsive $A^1\Pi$ state.

I distinguish now a different kind of predissociation which can take place due to strong interaction between states of the same symmetry and multiplicity. In *coupled states photodissociation* a first-order perturbation treatment is no longer appropriate. The adiabatic Born-Oppenheimer states, (one attractive, one repulsive) undergo an avoided crossing as shown in Figure 1c), coupling strongly through the radial component of the nuclear kinetic energy operator. The nuclear wave functions $\chi_{jk'}$ of the coupled excited states are obtained as solutions of a set of coupled differential equations which include the off-diagonal electronic elements of d/dR. The photodissociation cross section for absorption into the coupled excited states can be written[3]

$$\sigma = C\ \Delta E_{k'v''} \left| \sum_j < \chi_{jk'}(R) \left| D_{ji}(R) \right| \chi_{iv''}(R) > \right|^2 \qquad (3)$$

where C is a constant, and $D_{ji}(R)$ is the electronic dipole transition moment.

Obtaining an accurate description of several states of the same symmetry is difficult. Evaluation of the d/dR coupling matrix elements is also not trivial. Use of the Hellman-Feynman theorem allows one to express this matrix element in terms of the one-electron matrix element z_{iA}/r_{iA}^3 in which z_{iA} and r_{iA} are electron coordinates with respect to nucleus A,[5] but reliable values can usually be obtained only when nucleus A is H or He. Direct numerical differentiation has also been used, and very recently, analytic gradient techniques[6] have been used.

In OH, van Dishoeck et al.[7] have shown that an efficient dissociation pathway is provided through the $3^2\Pi$ state, which undergoes a strong avoided crossing with the $2^2\Pi$ state around 2.2 a_o. The absorption feature in this case consists of a series of resonances with Beutler-Fano line profiles superposed on a smooth background cross section.

DISSOCIATIVE RECOMBINATION

There are both direct and indirect mechanisms which may contribute to the dissociative recombination process. The indirect mechanism, which I will not mention further, occurs when a molecular ion captures an electron into a vibrationally excited Rydberg state which is then dissociated by repulsive states of the neutral molecule. The direct mechanism, as shown in Figure 2, involves capture of an electron of energy ε into a dissociative resonance state, AB** of the neutral system. The overall process is an important destruction pathway for molecular ions in the earth's atmosphere and interstellar clouds. Heating occurs as neutral species are created with significant kinetic energy.

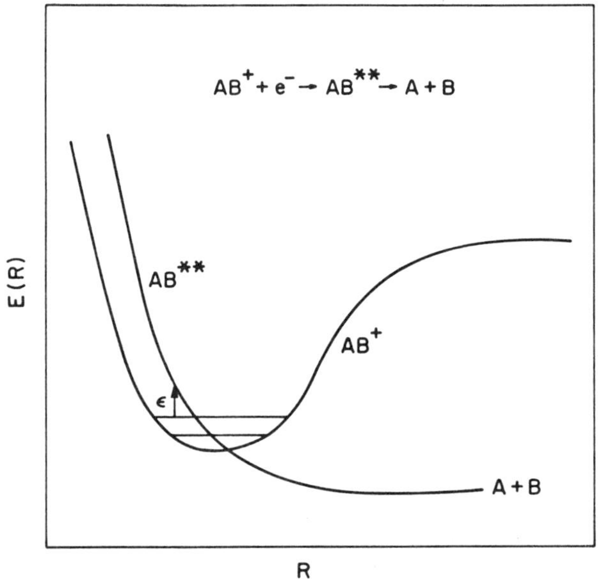

FIGURE 2. Potential curves relevant to the direct dissociative recombination of molecular ion AB^+.

The cross section for capture from vibrational level v of the molecular ion can be written:

$$\sigma \; \alpha \; |\langle \chi_v(R) | \Gamma^{\frac{1}{2}}(R) | \chi_k \rangle|^2 \tag{4}$$

in which χ_v and χ_k are the bound and continuum vibrational functions for the molecular ion AB^+ and the resonance state AB**, respectively. The width can be written[8]

$$\Gamma(R) = 2\pi |\langle \{\psi_{ion}(\underline{r}, R)\phi(\varepsilon)\} | H_{el} | \Psi_{res}(\underline{r}, R)\rangle|^2 \tag{5}$$

in which $\{\psi_{ion}(\underline{r}, R)\ \phi(\varepsilon)\}$ represents the antisymmetrized product of a continuum function for the electron and an electronic wavefunction for the molecular ion. The electronic wavefunction for the dissociating state is written $\Psi_{res}(\underline{r}, R)$ and the coupling operator above is the electronic Hamiltonian.

There a number of reasons that an accurate description of this process is difficult. One needs a good description of the incoming electron in the field of the molecular ion. Evaluation of the width is complicated by the presence of the electron continuum function and conventional molecular integral programs and methods are not usually useful. The resonance state needed is a doubly-excited diabatic valence state which cannot include Rydberg character or else the desired crossing would never occur. It is very tricky to obtain such states in ab initio calculations. The crossing of the neutral dissociative state and the ion curve must be described accurately because of the extreme sensitivity of the cross section to the overlaps of the vibrational functions.

Guberman[8] has addressed many of these problems in the case of O_2^+ and has calculated rates for production of $O(^1D)$ and $O(^1S)$ through this dissociative recombination process. He has used high Rydberg orbitals to represent the free electron. With such orbitals the short-range behavior of the continuum electron is described well and the long-range behavior which is not correct does not contribute to the integral in equation (5). The advantage in using Rydbergs is that the matrix elements in (5) can be evaluated by conventional molecular structure codes.

CHARGE TRANSFER

Charge transfer is an important process in plasmas, from the low density plasmas found in certain astrophysical environments to the high temperature and density plasmas created in fusion devices. At low energies it is particularly appropriate to use adiabatic molecular states of the aggregate molecular system. The charge transfer essentially takes place at the avoided crossings of the adiabatic potential curves where the coupling matrix elements of the nuclear kinetic energy and momentum are large.

The nuclear wavefunctions describing the scattering may be expanded in terms of partial waves which satisfy a set of coupled differential equations[5] involving the relevant potential energy curves and coupling matrix elements. For an accurate description of this process, complete potential curves which dissociate correctly and exhibit avoided crossings at the correct internuclear separation are needed. This requires that the one-particle as well as n-particle basis set in the electronic wavefunction be capable of describing the ionization of each atomic species. Another difficulty, yet again, is the accurate calculation of the nuclear radial coupling matrix elements.

A simple example of the charge transfer process has been treated by Kirby and Heil[9] for the AlH^{+3} system:

$$Al^{+3}(^1S) + H \rightarrow Al^{+2}(\ldots 3s,\ ^2S) + H^+$$
$$\rightarrow Al^{+2}(\ldots 3p,\ ^2P) + H^+$$
$$Al^{+2}(\ldots 3d,\ ^2D) + H^+$$

Four $^2\Sigma^+$ potential curves arising from the various separated atom limits must be computed in an unbiased fashion--a task made easier in this case because of the lack of huge correlation energy differences in the asymptotes.

SOME RECENT ADVANCES IN AB INITIO METHODS

In examining three very different molecular processes, I have indicated some of the challenges posed in an accurate ab initio treatment of these problems. There is currently work on-going in the quantum chemistry/molecular structure community which addresses some of these issues. Many advances have only been possible because of the availability of supercomputers.

Saxe, Schaefer and Handy[10] were among the first to use a full configuration interaction calculation to evaluate the limitations and successes of truncated CI methods. This work has been extended by Langhoff, Bauschlicher, Taylor and others[11] in their studies of the convergence of the energy surface, dipole moments, and dipole transition moments with respect to one-particle basis sets as well as the n-particle space (i.e., the level of CI in the wavefunction). They have also carried out full-CI benchmark calculations on multiple roots of the same symmetry and have evaluated methods to describe both singly- and multiply-bonded systems. This work is of critical importance in understanding what is required to obtain highly accurate ab initio results. New molecular codes are becoming available which accurately compute operators and coupling matrix elements.[6] These developments, which to a great extent go hand in hand with supercomputing capabilites will have a large impact on our ability to describe quantitatively many processes involving a wide variety of molecular systems in the future.

REFERENCES

1. E. F. van Dishoeck, IAU Symposium 120, Astrochemistry, edited by M. S. Vardya and S. P. Tarafdar (Reidel, Dordrecht), p. 51-65.
2. K. Kirby, R. P. Saxon and B. Liu, Astrophys. J., 231, 637 (1979).
3. K. Kirby and E. F. van Dishoeck, to be published in Advances in Atomic and Molecular Physics, Volume 24, (Academic Press, 1988).
4. E. F. van Dishoeck, M. C. van Hemert, and A. Dalgarno, J. Chem. Phys. 77, 3693 (1982).
5. T. G. Heil, S.E. Butler and A. Dalgarno, Phys. Rev. A, 23, 1100 (1981).
6. B. H. Lengsfield, P. Saxe and D.R. Yarkony, J. Chem. Phys. 81, 4549 (1984); P. Saxe, B. H. Lengsfield and D.R. Yarkony, Chem. Phys. Lett., 113, 139 (1985); B.H. Lengsfield and D.R. Yarkony, J. Chem. Phys., 84, 348 (1986); P. Saxe and D. R. Yarkony, J. Chem. Phys., 86, 321 (1987).
7. E. F. van Dishoeck, M. C. van Hemert, A. C. Allision and A. Dalgarno, J. Chem. Phys. 81, 5709 (1984).
8. S. L. Guberman, Planet. Space Sci., 36, 47 (1988); S. L. Guberman, Nature, 327, 408 (1987).
9. K. Kirby and T. G. Heil, J. Chem. Phys. 85, 6420 (1986).
10. P. Saxe, H. F. Schaefer and N.C. Handy, Chem. Phys. Lett. 79, 202 (1981).
11. C. W. Bauschlicher and P. R. Taylor, J. Chem. Phys., 86, 2844 (1987); C. W. Bauschlicher and S. R. Langhoff, J. Chem. Phys. 86, 5595 (1987); S. R. Langhoff, C. W. Bauschlicher and P. R. Taylor, J. Chem. Phys., 86, 6992 (1987), C.W. Bauschlicher and S. R. Langhoff, J. Chem. Phys., 87, 4665 (1987).

NONPERTURBATIVE APPROACHES TO ATOMIC AND MOLECULAR MULTI-PHOTON PROCESSES IN INTENSE LASER FIELDS

SHIH-I CHU
Dept. of Chemistry, Univ. of Kansas, Lawrence, KS 66045 U.S.A.

Abstract Several generalized semiclassical Floquet formalisms and quasi-energy methods recently developed for *ab initio* nonperturbative treatment of the nonlinear dynamical behavior of atoms and molecules in intense periodic or multi-color laser fields are presented. Applications to intense field multiphoton excitation, ionization, and dissociation as well as to intensity-dependent nonlinear optical processes are discussed.

INTRODUCTION

The study of the behavior of atoms and molecules in intense laser fields is among the most active fields in modern physics, encompassing a wide spectrum of problems from basic to very applied.[1] Many features make such problems different from traditional *weak* interactions of radiation with atoms and molecules. First of all, the interactions are *nonlinear*, which brings in a number of distinctly new high-order multiphoton phenomena. Second, a multiphoton transiton (i.e. absorption of many photons by a single atom or molecule) probes atomic and molecular structure in a more delicate and detailed way than in single-photon spectroscopy. Third, owing to the high intensities of lasers available, one can access high levels of excitation, thus introducing qualitatively new chemistry and physics.

Although the end result of such interactions is evident, the underlying dynamics and interplay between atomic/molecular structure and intense radiation is far from clear and difficult to explore. The main obstacles stem from the extreme complexity of the atomic and molecular level structure and high-order laser-atom (molecule) interactions. Indeed the current understanding of multiphoton processes has been largely confined to the phenomenological level and lacks quantitative basis. Furthermore, the conventional perturbative methods are applicable only to weak field and non-resonant problems and are difficult to extend to intense field and multiphoton resonance phenomena.

The theory of multiphoton processes can be formulated in a fully quantum mechanical or semiclassical fashion. In the former approach, both the system and the field are treated quantum mechanically, while in the semiclassical approach, the system is described by a time-dependent Schrodinger equation in which the effect of the radiation field is represented by an effective Hamiltonian consistent with Maxwell's equations. The fully quantized methods are generally difficult to apply as here one is dealing with a large number of time-independent coupled 2nd order differential equations subject to appropriate asymptotic boundary conditions. In contrast, only first-order initial value problems are involved in semiclassical methods. The semiclassical approach usually leads to Hamiltonians periodic in time through an assumed sinusoidal time variation of the electromagnetic field. The use of Floquet theorem[2] can further greatly simplify the dynamical study. In the following sections, I shall focus on the recent developments of the semiclassical Floquet methods for the treatment of intense field multiphoton and nonlinear optical processes. First, let me discuss how the Floquet theorem may be implemented in the conventional semiclassical treatment of multiphoton processes in periodic fields.

CONVENTIONAL FLOQUET THEORETICAL APPROACHES

There are two ways to implement the Floquet theorem[2] into numerical calculations. The first method is the time-dependent propagation technique in which one numerically intergrates the Schrodinger equation ($\hbar=1$)

$$i\partial_t\hat{U}(t,t_0) = \hat{H}(t)\hat{U}(t,t_0), \tag{1}$$

where the Hamiltonian is periodic in time, namely, $\hat{H}(t+T) = \hat{H}(t)$ with $T = 2\pi/\omega$ being the period, and $\hat{U}(t,t_0)$ is the time-evolution operator defined as

$$\psi(t) = \hat{U}(t,t_0)\psi(t_0), \text{ with } \hat{U}(t_0,t_0) = 1. \tag{2}$$

Floquet theorem[2] asserts that

$$\hat{U}(t,t_0) = \hat{P}(t,t_0) \exp[-i\hat{Q}(t-t_0)], \tag{3}$$

where $\hat{Q}$ is the time-independent quasi-energy operator, and $\hat{P}(t,t_0) = \hat{P}(t+T,t_0)$ is periodic. Since $\hat{U}(t_0+nT,t_0) = [\hat{U}(t_0+T,t_0)]^n$, one sees that the time evolution operator over one period, $\hat{U}(t_0+T,t_0)$, provides essentially all the information we ever need about the long-time behavior of the system. Further, in the truncated basis, $\hat{U}(t_0+T,t_0)$ may be diagonalized by some unitary transformation S,

$$S^{+}\hat{U}(t_0+T,t_0)S = \exp(-iD), \tag{4}$$

where D is a diagonal matrix. Thus

$$\hat{U}(t_0+T,t_0) = Se^{-iD}S^{+} \tag{5}$$

where upon

$$\hat{U}(t_0+nT,t_0) = Se^{-inD}S^{+}. \tag{6}$$

The second way to implement the Floquet theorem is via the so-called <u>time-indpendent Floquet Hamiltonian</u> method.[3-5] In this approach, one transforms the periodically time-dependent Hamiltonian $\hat{H}(t)$ into an equivalent <u>time-independent</u> infinite-dimensional Floquet matrix eigenvalue problem. As an example, consider the multiphoton excitation of vibrational-rotational states of a diatomic molecule under the periodic perturbation

$$V(t) = -\vec{\mu}(r) \cdot \vec{E}_0 \cos(\omega t+\delta), \tag{7}$$

where μ is the dipole moment operator of the molecule, E_0, ω, and δ are the field amplitude, frequency, and phase, respectively. The corresponding time-independent Floquet matrix $\hat{H}_F$ posseses a block tridiagonal form as shown in Fig. 1. The determination of the vibration-rotational quasi-energies and quasi-energy states (QES) thus reduces to the solution of a time-independent

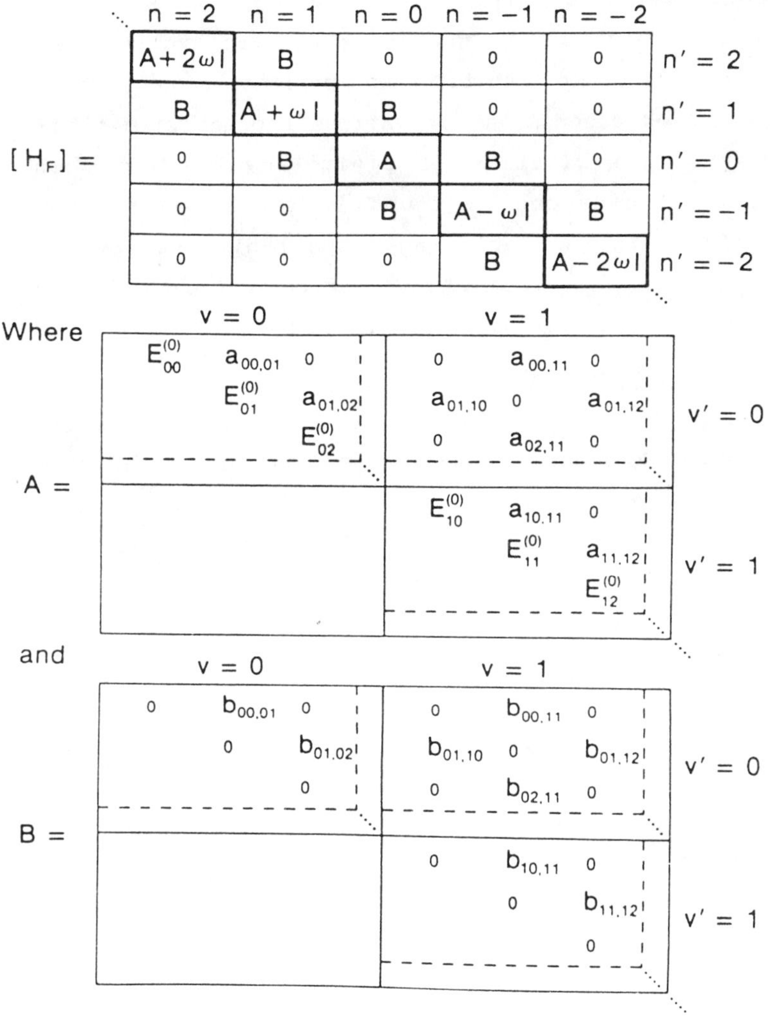

Fig.1 Structure of Floquet Hamiltonian H_F for MPE of molecules. E_{vj} are unperturbed vibrational-rotational energies and $b_{vj,\, v'j'}$ are electric dipole coupling matrix elements.

Floquet matrix eigenproblem. Fig. 1 shows that $\hat{H}_F$ has a periodic structure with only the number of ω's in the diagonal elements varying from block to block. This structure endows the quasi-energy eigenvalues and eigenvectors of $\hat{H}_F$ with important periodic properties.[3-5]

Detailed discussions of various Floquet studies of multiphoton excitation of finite-level systems in periodic fields can be found in the recent review article.[5] In the following I shall discuss several major problems encountered by these conventional Floquet techniques when applied to realistic systems.

Problems with Conventional Floquet Methods

The first is the large atom (molecule)-field basis set problem. Consider the dimensionality of the Floquet matrix such as Fig. 1 for example. In practical calculations, one truncates the Floquet matrix into N by N in dimension where $N = N_F N_V N_J$, where N_F is the number of Floquet photon blocks, N_V the number of vibrational levels included in one Floquet block, and N_J the number of rotational states included in one vibrational level. As N increases rapidly with the size of the molecule and with the order of multiphoton transitions, and as the computational expense grows as N^3, the full Floquet analysis can become prohibitively costly even for simple molecules. While the use of a supercomputer can provide solutions to some of these problems, the development of new approximation techniques capable of providing accurate results yet involving much smaller Floquet matrix manipulation is in great demand.

The second general limitation of conventional Floquet techniques is that they are applicable only to monochromatic (i.e. one-laser) field problems. However, many recent experiments involve the use of more than one laser field, namely, multi-color or multi-frequency excitation in polychromatic fields. In such cases, the Hamiltonians are no longer periodic in time and the Floquet theorem is simply not valid.

The third general problem arises from the fact that conventional Floquet techniques are based on the use of the Schrodinger equation and classical fields. They cannot therefore be applied to the study of nonlinear optical processes where spontaneous decays and collisional dampings play an essential role.

Finally, the conventional Floquet methods with Hermitian Hamiltonian can only deal with bound-bound multiphoton excitation (MPE) processes, but not bound-free multiphoton ionization (MPI) or multiphoton dissociation (MPD) problems.

In the following section, I shall discuss several generalized Floquet formalisms recently developed in our laboratory which offer possible solutions to these problems.

GENERALIZED FLOQUET FORMALISMS

Fig. 2 shows an outline of several generalized Floquet methods and atomic and molecular processes amenable to the generalized treatments. As detailed descriptions of these methods have already been given in recent review articles,[5,6] I shall discuss below only the essential

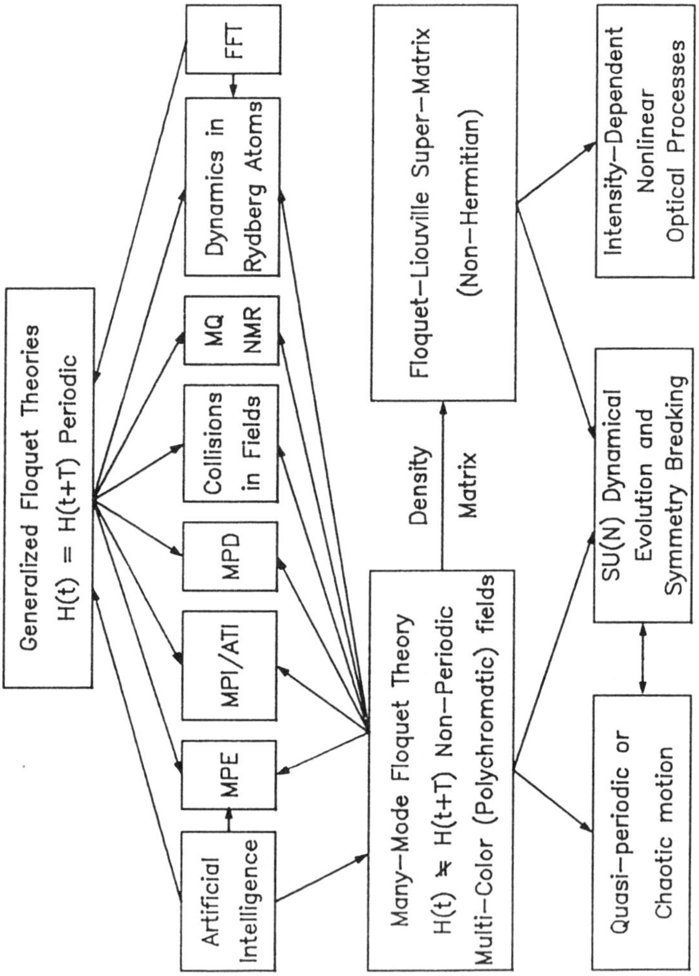
Generalized Floquet Theories
H(t) = H(t+T) Periodic
Artificial Intelligence
MPE
MPI/ATI
MPD
Collisions in Fields
MQ NMR
Dynamics in Rydberg Atoms
FFT
Many-Mode Floquet Theory
H(t) ≠ H(t+T) Non-Periodic
Multi-Color (Polychromatic) fields
Density Matrix
Floquet-Liouville Super-Matrix
(Non-Hermitian)
Quasi-periodic or Chaotic motion
SU(N) Dynamical Evolution and Symmetry Breaking
Intensity-Dependent Nonlinear Optical Processes

Fig.2

physical contents of the formalisms but not the technical details.

Generalized Floquet Methods for Multiphoton Processes in Periodic Fields

Several methods for effectively reducing the large dimensionality of Floquet matrix into manageable size have been developed.[5] The methods have been applied to the studies of multiphoton excitation (MPE) of atoms and molecules,[7] to laser-induced charge-exchange reactions,[8] to multiple-quantum (MQ) NMR dynamics in coupled-spin systems such as oriented benzene molecules,[9] and to multiphoton dynamics and quantum diffusion in Rydberg atoms[10] etc.

One such method is termed the most probable path approach (MPPA) first introduced by Tietz and Chu[11] in an *ab initio* study of high-order MPE of SO_2. A brute-force attempt to calculate polyatomic MPE would soon become prohibitive due to the large number of vibrational-rotational states and the huge size of the Floquet matrix needed for achieving convergence. For typical 15-photon calculations for SO_2 including only vibrational ladders, for example, a matrix on the order of 10,000 by 10,000 would have to be diagonalized at each frequency and field strength. In any exact Floquet calculation, however, the majority of the molecule-field states are unimportant due to large detuning or very small coupling matrix elements. The most probable path approach is a practical strategy introduced to determine which molecule-field states are, in fact, important at each step of the multiphoton processes. The procedure is derived from algorithms which utilize artificial intelligence (AI) to prune the number of choices at each node (photon order) of a decision tree.[12] Similar to some minimax game playing programs, the MPPA examines the possible paths to take at each photon order iteration with the static evaluation function given by Nth order perturbation theory (this is a breadth-first search). If all paths were followed exhaustively, the problem would be beyond practical solution. In game theory, one answer is to ignore paths which start with very poor moves. The MPPA likewise uses a breadth-limiting heuristic technique and discards any paths for which the Nth order coupling term is small (with respect to other Nth order terms). The reduction of the basis set is quite dramatic and leads to many-orders-of-magnitude savings in computer time, yet maintains good accuracy in most cases. The MPPA-AI technique has recently been further refined and applied to the study of MPE and three-dimensional quantum diffusion phenomena in Rydberg H atom (with initial principal quantum number $n_0 \geq 60$) driven by microwave fields.[11] In the latter case, the MPPA allows the reduction of an ultralarge Floquet matrix (on the order of millions) to an effective matrix of a manageable size (on the order of several hundreds).

Non-Hermitian Floquet Hamiltonian Methods for MPI/MPD/ATI

As mentioned above, one of the major problems in conventional Floquet techniques lies in its limitation to only bound-bound processes where the Floquet Hamiltonian $\hat{H}_F$ is Hermitian and the quasi-energy is real. To treat bound-free MPI/MPD processes, analytical continuaiton of $\hat{H}_F$ to non-Hermitian Hamiltonian is necessary. The most expedient way is to invoke the use of the complex scaling technique,[13] $\vec{r} \rightarrow \vec{r}e^{i\alpha}$, to the Schrodinger equation,

$$i\partial_t \psi(\vec{r}e^{i\alpha},t) = \hat{H}(\vec{r}e^{i\alpha},t)\psi(\vec{r}e^{i\alpha},t), \quad (8)$$

where $\hat{H}(\vec{r}e^{i\alpha},t)$ is now a non-Hermitian periodic Hamiltonian. The wavefunction $\psi(\vec{r}e^{i\alpha},t)$, can be written, according to the Floquet theorem,[2]

$$\psi(\vec{r}e^{i\alpha},t) = \exp(-i\varepsilon t)\ \Phi(\vec{r}e^{i\alpha},t), \quad (9)$$

where ε is the <u>complex</u> quasi-energy, and the function $\Phi(\vec{r}e^{i\alpha},t)$ is periodic in time. Following the procedure of the time-independent Floquet Hamiltonian method,[5] Eq. (8) can be converted into a non-Hermitian Floquet matrix $\hat{H}_F(\alpha)$ eigenvalue problem. The complex scaling transformation distorts the continuous spectrum away from the real axis, exposing the Stark resonances (i.e. complex quasi-energy states) in appropriate higher Riemann sheets, and also allowing use of finite variational expansions employing the L^2 basis function chosen from a complete discrete basis. The use of a complete L^2 basis obviates the necessity of explicit introduction of exact atomic or molecular bound and continuum states, thus reducing all computations to those involving finite-dimensional non-Hermitian matrices. The use of complex coordinates not only allows direct calculation of eigenvalue parameters associated with complex dressed states, but completely avoids numerical problems arising from strong coupling between overlapping atomic or molecular continua. The real part of the complex quasi-energy (E_R) provides ac-Stark shifted energy, whereas the imaginary part ($\Gamma/2$) determines the total MPI or MPD rate. The methods have been applied to the study of intensity-dependent MPI of atom hydrogen in linearly[14] and circularly[15] polarized fields, to photoionization[16] in superstrong magnetic fields, and to two-photon dissociation of H_2^+ and HD^+ from highly excited vibrational states.[17] For a review of these complex quasi-energy developments and other related MPI/MPD methods, the reader is referred to ref. 5 for details. More recently, the non-Hermitian Floquet matrix technique has been extended to the study of multiphoton "above-threshold ionization" (ATI) of atoms in very intense laser fields.[18] Finally, an alternative time-dependent technique involving the use of the Fast Fourier transformation (FFT) technique and the Floquet theorem for the investigation of the dynamical evolution of MPI/MPD/ATI processes is being pursued in our laboratory.[19]

Many-Mode Floquet Theory for Multiphoton Processes in Multi-Color or Polychromatic Fields

The Floquet methods described so far require the Hamiltonian to be explicitly periodic in time and are therefore applicable to problems involving only one monochromatic radiation field. Exact extension of the one-mode Floquet Hamiltonian formalism[3,4] to a generalized many-mode Floquet theory[5,20] (MMFT) was found in 1983. This makes it possible to treat, for the first time, the time-dependent problem of any finite-level system exposed to polychromatic fields as an equivalent time-independent infinite-dimensional eigenvalue problem.

Without loss of generality, let us consider the interaction of an arbitrary N-level system with two linearly polarized monochromatic fields. The Schrodinger equation can be written as ($\hbar = 1$)

$$\mathcal{H}(\vec{r},t)\psi(\vec{r},t) = 0, \qquad \mathcal{H}(\vec{r},t) = H(\vec{r},t) - i\partial/\partial t$$

where

$$H(\vec{r},t) = H_0(\vec{r}) + \sum_i V_i(\vec{r},t) \tag{10}$$

and

$$V_i(\vec{r},t) = -\vec{\mu} \cdot \vec{E}_i \cos(\omega_i t+\phi_i), \qquad i = 1,2. \tag{11}$$

H_0 is the unperturbed Hamiltonian with orthonormal eigenstates $\{|\alpha\rangle, |\beta\rangle,\ldots\}$ and eigenvalues $\{E_\alpha, E_\beta,\ldots\}$, μ is the dipole moment operator, E_i, ω_i, and ϕ_i are, respectively, the (peak) amplitude, the frequency, and the initial phase of the ith monochromatic field ($i=1,2$). For the Hermitian operator $\mathcal{H}(\vec{r},t)$, one can now introduce the composite Hilbert space $R + T_1 + T_2$. The spatial part R is spanned by $\{|\alpha\rangle, |\beta\rangle,\ldots\}$, and, in the two-mode temporal part, T_1 is spanned by $\exp(in_1\omega_1 t)$ and T_2 by $\exp(in_2\omega_2 t)$, where $n_1, n_2 = 0, \pm 1, \pm 2,\ldots$. In the two-mode Floquet theory, the time-evolution operator in its matrix form is shown to be

$$\hat{U}_{\gamma_1\gamma_2}(t,t_0) \equiv \langle\gamma_1|\hat{U}(t,t_0)|\gamma_2\rangle,$$

$$= \sum_{n_1}\sum_{n_2} \langle\gamma_1 n_1 n_2|\exp[-i\hat{H}_F(t-t_0)]|\gamma_2 00\rangle$$

$$\times \exp[i(n_1\omega_1 + n_2\omega_2)t]. \tag{12}$$

Here $\hat{H}_F$ is the time-independent two-mode Floquet Hamiltonian defined in the generalized Floquet state basis,

$$|\gamma n_1 n_2\rangle \equiv |\gamma\rangle \times |n_1\rangle \times |n_2\rangle$$

and satisfies the infinite-dimensional engenvalue equation,

$$\sum_{\gamma_2 k_1 k_2} \langle\gamma_1 n_1 n_2|\hat{H}_F|\gamma_2 k_1 k_2\rangle\langle\gamma_2 k_1 k_2|\lambda\rangle = \lambda \langle\gamma_1 n_1 n_2|\lambda\rangle. \tag{13}$$

Fig. 3 depicts the structure of the two-mode Floquet Hamiltonian. The components are ordered in such a way that γ runs over unperturbed states (denoted by Greek letters) of H_0 before each change in n_1 and n_1, in turn, runs over before n_2. The quasi-energy eigenvalues $\{\lambda_{\gamma n_1 n_2}\}$ and their corresponding eigenvectors $\{|\lambda_{\gamma n_1 n_2}\rangle\}$ of $\hat{H}_F$ have the following useful periodicity properties, namely,

$$\lambda_{\gamma n_1 n_2} = \lambda_{\gamma 00} + n_1\omega_1 + n_2\omega_2, \tag{14}$$

and

$$\langle\gamma_1, n_1+q_1, n_2+q_2|\lambda_{\gamma_2, n_1+q_1, n_2+q_2}\rangle$$

$$= \langle\gamma_1 n_1 n_2|\lambda_{\gamma_2 n_1 n_2}\rangle. \tag{15}$$

The transition probability averaged over the initial

$$
[H_F] = \begin{bmatrix} \ddots & & & & & & \\ & A + 2\omega_2 I & B & 0 & 0 & 0 & \\ & B & A + \omega_2 I & B & 0 & 0 & \\ & 0 & B & A & B & 0 & \\ & 0 & 0 & B & A - \omega_2 I & B & \\ & 0 & 0 & 0 & B & A - 2\omega_2 I & \\ & & & & & & \ddots \end{bmatrix}
$$

WHERE

$$
A = \begin{bmatrix} \ddots & & & & & & \\ & C + 2\omega_1 I & X & 0 & 0 & 0 & \\ & X & C + \omega_1 I & X & 0 & 0 & \\ & 0 & X & C & X & 0 & \\ & 0 & 0 & X & C - \omega_1 I & X & \\ & 0 & 0 & 0 & X & C - 2\omega_1 I & \\ & & & & & & \ddots \end{bmatrix}
$$

$$
C = \begin{bmatrix} E_\alpha & 0 & 0 & \\ & E_\beta & 0 & \\ & & E_\gamma & \\ & & & \ddots \end{bmatrix}
\qquad
X = \begin{bmatrix} 0 & V^{(1)}_{\alpha\beta} & V^{(1)}_{\alpha\gamma} & \\ & 0 & V^{(1)}_{\beta\gamma} & \\ & & 0 & \\ & & & \ddots \end{bmatrix}
$$

$$
B = \left[\begin{array}{c|c} \begin{matrix} 0 & V^{(2)}_{\alpha\beta} & V^{(2)}_{\alpha\gamma} & \\ & 0 & V^{(2)}_{\beta\gamma} & \\ & & 0 & \\ & & & \ddots \end{matrix} & \mathrm{O} \\ \hline \mathrm{O} & \begin{matrix} 0 & V^{(2)}_{\alpha\beta} & V^{(2)}_{\alpha\gamma} & \\ & 0 & V^{(2)}_{\beta\gamma} & \\ & & 0 & \\ & & & \ddots \end{matrix} \end{array}\right] \ddots
$$

Fig.3 Structure of two-mode Floquet Hamiltonian. It can be generalized to N-field problem.

time t_0 while keeping the elapsed time $t-t_0$ fixed is given by

$$P_{\alpha\to\beta}(t-t_0) = \sum_{k_1 k_2} |\langle\beta k_1 k_2|\exp[-i\hat{H}_F(t-t_0)]|\alpha 00\rangle|^2. \quad (16)$$

Performing the long-time average over $t-t_0$ gives the time averaged transition probability

$$\bar{P}_{\alpha\to\beta} = \sum_{k_1 k_2} \sum_{\gamma \ell_1 \ell_2} |\langle\beta k_1 k_2|\lambda_{\gamma\ell_1\ell_2}\rangle\langle\lambda_{\gamma\ell_1\ell_2}|\alpha 00\rangle|^2. \quad (17)$$

The MMFT has been applied to the study of two-color MPE of atoms and molecules,[20,21] to the study of SU(N) dynamical evolution and symmetry breaking in N-level systems driven by polychromatic fields,[22] and to the study of quasi-periodic or chaotic behavior of quantum systems under the perturbation of strong multi-color fields.[23]

Floquet-Liouville Super-Matrix Formalism for Intensity-Dependent Nonlinear Optical Processes

The various Flouqet approaches discussed above are all based on the solution of the wave function (or time propagator) of the Schrodinger equation and are best applied to the studies of intense field MPE/MPI/MPD processes in which relaxation processes such as spontaneous decays and collisional dampings can be ignored. However, in nonlinear optical processes such as resonance fluorescence, multiple wave mixings, and Raman scattering etc., relaxations often play a significant role.[24] The correct starting point is to consider the Liouville equation for the time development of the density matrix operator of atoms and molecules, allowing for a more natural introduction of the damping mechanisms.

The Liouville equation for the time evolution of a set of N-level quantum systems, interacting with several coherent linearly polarized monochromatic fields, undergoing relaxation by Markovian processes[24] is ($\hbar$=1)

$$i\partial_t\rho(t) = [\hat{H}(t),\rho(t)] + i[\hat{R},\rho(t)]. \tag{18}$$

Here ρ is the density matrix of ths system, reduced by an averaging over all irrelevant degrees of freedom acting as a thermal bath, and $\hat{H}(t)=\hat{H}_0 + \hat{V}(t)$. $\hat{H}_0$ is the unperturbed Hamiltonian with eigenvalues $\{E_\alpha\}$ and elgenvectors $\{|\alpha\rangle\}$, $\alpha=0,1,2,\ldots,N-1$, and $V(t)$ is the interaction Hamiltonian between the system and the M-mode classical fields given by

$$V(t) = -\sum_{i=1}^{M} \vec{\mu}\cdot\vec{\varepsilon}_i \cos(\omega_i t + \phi_i). \tag{19}$$

The relaxation term $[\hat{R},\rho(t)]$ consists of T_1 (population damping) and T_2 (coherent damping) mechanisms which are due to the coupling of the system to the thermal bath by radiative decays and collisional relaxations etc. It has recently been shown that the polychromatic time-dependent Liouville equation, Eq. (18), can be exactly transformed into an equivalent time-independent infinite-dimensional <u>non-Hermitian</u> super-eigenvalue equation,[25] namely,

$\hat{L}_F =$ (the pattern continues beyond the matrix in both directions, shown by dots ⋱ at the top-left and bottom-right)

$A+2\omega_2 I$	B	0	0	0
B^*	$A+\omega_2 I$	B	0	0
0	B^*	A	B	0
0	0	B^*	$A-\omega_2 I$	B
0	0	0	B^*	$A-2\omega_2 I$

WHERE

$A =$ (the pattern continues beyond the matrix in both directions, shown by dots ⋱)

$C+2\omega_1 I$	X	0	0	0
X^*	$C+\omega_1 I$	X	0	0
0	X^*	C	X	0
0	0	X^*	$C-\omega_1 I$	X
0	0	0	X^*	$C-2\omega_1 I$

$B =$ (the pattern continues beyond the matrix in both directions, shown by dots ⋱)

Y	0	0	0	0
0	Y	0	0	0
0	0	Y	0	0
0	0	0	Y	0
0	0	0	0	Y

$C =$

$-i(\gamma_{ab}+\gamma_{ba})$	0	0	0
$i\gamma_{ab}$	$-i\gamma_{ba}$	0	0
0	0	$-\omega_{ba}-i\Gamma_{ba}$	0
0	0	0	$\omega_{ba}-i\Gamma_{ba}$

AND

$X =$

0	0	$-V_{ba}^{(1)}$	$V_{ab}^{(1)}$
0	0	$V_{ba}^{(1)}$	$-V_{ab}^{(1)}$
$-V_{ab}^{(1)}$	$V_{ab}^{(1)}$	0	0
$V_{ba}^{(1)}$	$-V_{ba}^{(1)}$	0	0

$Y =$

0	0	$-V_{ba}^{(2)}$	$V_{ab}^{(2)}$
0	0	$V_{ba}^{(2)}$	$-V_{ab}^{(2)}$
$-V_{ab}^{(2)}$	$V_{ab}^{(2)}$	0	0
$V_{ba}^{(2)}$	$-V_{ba}^{(2)}$	0	0

Fig. 4 Structure of the Floquet-Liouville supermatrix $\hat{L}_F$ for the case of two-level systems (with level spacing ω_{ba}) in linearly polarized bichromatic fields. ω_1 and ω_2 are the two radiation frequencies, $V_{ab}^{(i)}$ (i=1,2) are the electric dipole couplings, and γ_{ab}, γ_{ba}, and $\Gamma_{ba} \equiv (\gamma_{ab}+\gamma_{ba})/2$ are relaxation parameters.

$$\sum_{\sigma}\sum_{\tau}\sum_{k} \langle\alpha\beta;\{m\}|\hat{L}_F|\sigma\tau;\{k\}\rangle\langle\sigma\tau;\{k\}|\Omega_{\mu\nu;\{n\}}\rangle$$

$$= \Omega_{\mu\nu;\{n\}}\langle\alpha\beta;\{m\}|\Omega_{\mu\nu;\{n\}}\rangle, \qquad (20)$$

where $\hat{L}_F$ is the <u>time-independent</u> many-mode Floquet-Liouville superoperator defined in terms of the generalized tetradic-Floquet basis

$|\alpha\beta;\ \{m\}\rangle \equiv |\alpha\rangle\langle\beta|x|\{m\}\rangle$, with $\{m\} = m_1, m_2, \ldots, m_M$. The structure of the Floquet-Liouville super-matrix $\hat{L}_F$ is illustrated in Fig. 4 for the simplest two-level two-mode case.

The development of the Floquet-Liouville super-matrix (FLSM) formalism thus allows the study of intense field nonlinear optical processes in terms of a few complex super-eigenvalues and eigenvectors. This yields a numerically stable and computationally efficient approach for the unified treatment of nonresonant and resonant, one- and multiple-photon, steady-state, and transient phenomena in nonlinear optical processes, much beyond the conventional perturbative and rotating wave approximation (RWA) approaches. The FLSM method has been applied to the study of intense field multiphoton resonance fluorescence spectra,[25] intensity-dependent nonlinear optical susceptibilities,[26] and multiple-wave mixings[26] etc.

CONCLUSION

Several generalized Floquet formalisms recently developed are briefly reviewed. The methods are shown to provide powerful and practical nonperturbative techniques for ab initio comprehensive probe of a wide range of multiphoton and nonlinear optical processes in intense monochromatic or polychromatic laser fields.

ACKNOWLEDGEMENTS

The works described in this article were supported in part by the Department of Energy (Division of Chemical Sciences) and by the John Simon Guggenheim Fellowship.

REFERENCES

1. Recent reviews can be found, for example, in Coherent Nonlinear Optics, edited by M.S. Feld and V.S. Letokhov (Springer, New York, 1980); Photoselective Chemistry, edited by J. Jortner, R.D. Levine, and S.A. Rice, Adv. Chem. Phys., Vol. 47 (Wiley, New York, 1981); Advances in Multiphoton Processes and Spectroscopy, edited by S.H. Lin, Vols. 1-3 (World Scientific, 1984, 1985, 1987).
2. G. Floquet, Ann. de l'Ecole Norm. Sup. 12, 47 (1983); S.H. Autler and C.H. Townes, Phys. Rev. 100, 703 (1955).
3. J.H. Shirley, Phys. Rev. 138, B 979 (1965).
4. D.R. Dion and J.O. Hirschfelder, Adv. Chem. Phys. 35, 265 (1976).
5. S.I. Chu, Adv. At. Mol. Phys. 21, 197 (1985).
6. S.I. Chu, Adv. Chem. Phys. (1988).
7. See, for example, T.S. Ho and S.I. Chu, J. Chem. Phys. 79, 4708 (1983).
8. See, for example, T.S. Ho, C. Laughlin, and S.I. Chu, Phys. Rev. A 32, 122 (1985).

9. S.I. Chu and K. Wang, J. Chem. Phys. (1988).
10. K. Wang and S.I. Chu, Phys. Rev. A (1988).
11. J.V. Tietz and S.I. Chu, Chem. Phys. Lett. 101, 446 (1983).
12. See, for example, P.H. Winston, Artificial Intelligence (Addison-Wesley, Reading, MA 1979).
13. For recent reviews on the method of complex scaling, see (a) B.R. Junker, Adv. At. Mol. Phys. 18, 208 (1982); (b) W.P. Reinhardt, Ann. Rev. Phys. Chem. 33, 223 (1982).
14. S.I. Chu and W.P. Reinhardt, Phys. Rev. Lett. 39, 1195 (1977).
15. S.I. Chu, Chem. Phys. Lett. 54, 367 (1987).
16. S.K. Bhattacharya and S.I. Chu, J. Phys. B 18, L275 (1985).
17. C. Laughlin, K.K. Datta, and S.I. Chu, J. Chem. Phys. 85, 1403 (1986) and references therein.
18. S.I. Chu and J. Cooper, Phys. Rev. A 32, 2769 (1985); S.I. Chu and R.Y. Yin, J. Opt. Soc. Am. B4, 720 (1987).
19. T.F. Jiang and S.I. Chu, Phys. Rev. A (submitted).
20. T.S. Ho, S.I. Chu, and J.V. Tietz, Chem. Phys. Lett. 96, 464 (1983).
21. T.S. Ho and S.I. Chu, J. Phys. B 17, 2101 (1984).
22. T.S. Ho and S.I. Chu, Phys. Rev. A 31, 659 (1985); A 32, 377 (1985).
23. K. Wang and S.I. Chu, Phys. Rev. A (1988).
24. Y.R. Shen, The Principles of Nonlinear Optics (Wiley, New York, 1984).
25. T.S. Ho, K. Wang, and S.I. Chu, Phys. Rev. A 33, 1798 (1986).
26. K. Wang and S.I. Chu, J. Chem. Phys. 86, 3225 (1987).

THE NUCLEAR FEW-BODY SYSTEM

G. L. PAYNE
The University of Iowa, Iowa City, Iowa 52242, U.S.A.

Abstract Numerical techniques used to solve model Hamiltonians for nuclear few-body systems are reviewed, and a brief summary of the current calculations is presented. The use of these calculations to test the validity of the models by detailed comparisons of the numerical predictions with the experimental data is also discussed. The need for larger supercomputers to do the more detailed calculations is stressed, and a brief discussion of the possible calculations which could be accomplished with these computers is presented.

INTRODUCTION

The traditional approach of that branch of physics concerned with the study of atomic nuclei is to view the nucleus as a collection of nonrelativistic, pairwise-interacting nucleons. Thus, the nucleus is described by a model Hamiltonian which has realistic two-nucleon interactions, and one solves the nonrelativistic Schrödinger equation. Since this model neglects the subnuclear degrees of freedom such as mesons, it is clearly only an approximation to the true situation. That meson-exchange is responsible for the long-range nucleon-nucleon force is well known, and the presence of mesons within the nucleus must give rise to many-body forces. Also, the meson exchange currents produce important modifications in the electromagnetic properties of the nucleus. In addition, the nucleons and mesons are themselves composed of quarks, and the possibility that these degrees of freedom are reflected in nuclear properties is currently of considerable interest in nuclear physics. In order to use the atomic nucleus as a probe of quark degrees of freedom, it is important that we determine the validitity of the traditional model of the nucleus. Detailed comparisons between theoretical calculations and the experimental data serve as a test of the traditional approach. Any disagreements between theoretical predictions and experimental values provide an indication of the need for including subnuclear degrees of freedom. Only by determining where the traditional model fails can we utilize the nuclear system to test the theory of quantum chromodynamics.

The traditional model which provides a semiphenomenological description of nuclear structure and reactions has enjoyed considerable success. However, the inability to calculate observables accurately has been a serious impediment to the development of a more fundamental understanding of nuclear physics. During the past ten years there has been significant progress in the numerical calculations for the few-body problem. Much of this progress has been due to the availability of new supercomputers which now give us the capability of doing calculations which were impossible only a few years ago. The use of these computers along with the implementation of new algorithms which utilize the vector nature of these computers has made the few-body problem an important tool in nuclear physics. The three-body bound state can now be solved "exactly," that is, the model Hamiltonian can be solved to an arbitrary accuracy. Consequently, any discrepancies between the calculated values and the experimental values must be due to a failure of the model. This situation is considerably different from the previous calculations where it was difficult to determine whether the discrepancies were caused by the numerical approximations or by the model. The three-body scattering problem calculations have not reached the accuracy of the bound-state calculations, but rapid progress is being made and during the next few years these calculations will reach a level of accuracy equivalent to the bound-state calculations. In addition, accurate calculations for the four-body problem should become possible, especially as new and larger computers become available.

The few-body problem in nuclear physics encompasses many different systems and problems. For example, one can study the bound-state or scattering problem. These two problems require different numerical techniques and their solutions have fundamental differences. The bound-state problem has a unique solution for the Schrödinger equation, and many different numerical techniques exist which can be used the find the bound-state wave function. The uniqueness property is much more difficult for the scattering problem, and significant progress in the solution of the three-body scattering problem did not occur until after Faddeev's seminal work[1] on three-body scattering which reduced the problem to connected-kernel t-matrix equations. Equations with connected kernels can be solved using the well-known Fredholm theory, in contrast to the three-body Lippmann-Schwinger equations for which there is no unique solution.[1] The simplest three-body bound-state systems are the ^{3}H and ^{3}He nuclei. However, more complicated systems can often be approximated as three-body clusters. Examples of these cluster calculations include treating the ^{6}Li nucleus as an alpha-neutron-proton cluster or the ^{9}Be nucleus as an alpha-alpha-neutron cluster. For these cluster models there remains the question of the validity of the cluster approximation, and the numerical calculation can not be considered to be an "exact" calculation. Therefore, we limit our discussion to the trinucleon systems, that is, the ^{3}H and ^{3}He nuclei. The trinucleon scattering problem is either the neutron-deuteron or the proton-deuteron scattering problem, where both elastic scattering and three-body breakup can occur.

In the next section the numerical methods used to solve the three-body bound-state problem are reviewed, and the use of modern supercomputers to solve these these problems is discussed. Then the status of the the present calculations is presented. Next, the use of new and larger supercomputers for the scattering problem is discussed. Finally, the last section contains a brief summary of the status of the few-body problem in nuclear physics.

BOUND-STATE CALCULATIONS

As discussed in the previous section, we assume that the trinucleon system can be described by a nonrelativistic Hamiltonian of the form:

$$\begin{aligned} H &= H_0 + V(\mathbf{x}_1) + V(\mathbf{x}_2) + V(\mathbf{x}_3) + W(\mathbf{x}_1, \mathbf{x}_2, \mathbf{x}_3) \\ &= H_0 + V_1 + V_2 + V_3 + W, \end{aligned} \tag{1}$$

where H_0 is the free-particle kinetic energy operator, V_i is the two-body interaction between particles j and k, and $\mathbf{x}_i$ is the vector between particles j and k for i, j, k cyclic. The W is a possible three-body force which is discussed below, and it can be expressed as the sum of three terms. Therefore, we write

$$W = W_1 + W_2 + W_3. \tag{2}$$

Given the model Hamiltonian, the wave function for the trinucleon system is the solution to the three-body Schrödinger equation

$$(H - E)|\Psi\rangle = 0. \tag{3}$$

For the bound-state system one can solve the Schrödinger equation directly using a method such as the Rayleigh-Ritz Variational Principle.[2,3] Also, one could write the Schrödinger equation as a partial differential equation in configuration space and use a numerical procedure such as the hyperspherical expansion[4] to solve the equation. An alternate procedure[5] is to introduce the three Faddeev amplitudes and replace the Schrödinger equation by the three coupled Faddeev equations. Since most of the numerical calculations for the three-nucleon system use some form of the Faddeev equations, we first give a brief review of these equations for the bound-state problem.

Faddeev Equations

The three Faddeev equations, $|\Psi_i\rangle$, are defined by decomposing the total wave function into three parts by writing

$$|\Psi\rangle = \sum_{i=1}^{3} G_0(E)\tilde{V}_i|\Psi\rangle = \sum_{i=1}^{3} |\Psi_i\rangle, \tag{4}$$

where we have defined $\tilde{V}_i$ as $\tilde{V}_i = (V_i + W_i)$, and

$$G_0(E) = (E - H_0)^{-1} \tag{5}$$

is the free-particle resolvent. Introducing the resolvent operator

$$G_i(E) = (E - H_i)^{-1}, \tag{6}$$

where $H_i = H_0 + \tilde{V}_i$, the equation for $|\Psi_i\rangle$ can be written:

$$|\Psi_i\rangle = G_i(E)\tilde{V}_i[|\Psi_j\rangle + |\Psi_k\rangle] \tag{7}$$

for $i = 1, 2, 3$.

The three coupled equations for the $|\Psi_i\rangle$ comprise the Faddeev equations, and the solution of these equations is equivalent to the solution of the Schrödinger equation. To solve these equations one must choose a particular representation, and different groups have chosen different representations. Possible choices are the momentum-space representation,[6,7] the coordinate-space representation,[8,9] and a combination of the two representations.[10] In both the momentum-space and the coordinate-space representations, it is customary to use different coordinates for each of the three Faddeev amplitudes. In coordinate space one normally uses the three sets of Jacobi coordinates

$$\begin{aligned} \mathbf{x}_i &= \mathbf{r}_j - \mathbf{r}_k \\ \mathbf{y}_i &= \tfrac{1}{2}(\mathbf{r}_j + \mathbf{r}_k) - \mathbf{r}_i, \end{aligned} \tag{8}$$

where the $\mathbf{r}_i$ are the positions of the three nucleons, and i, j, k are taken cyclically. An example of these coordinates is shown in Figure 1. In

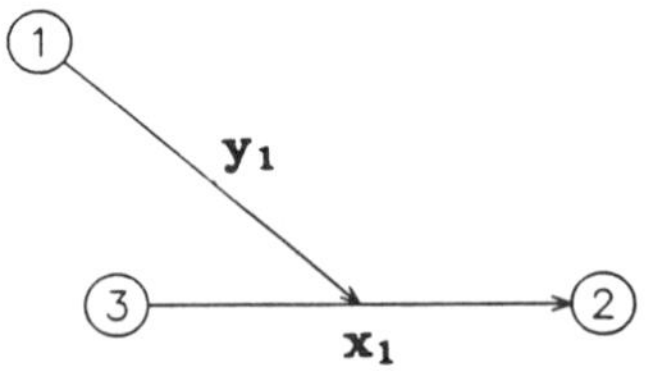

FIGURE 1. Vectors $\mathbf{x}_1$ and $\mathbf{y}_1$ for the three-nucleon system.

momentum space the relative momenta conjugate to the Jacobi coordinates are

$$\mathbf{p}_i = \tfrac{1}{2}\left(\mathbf{k}_j - \mathbf{k}_k\right)$$
$$\mathbf{q}_i = \tfrac{2}{3}\left[\tfrac{1}{2}\left(\mathbf{k}_j + \mathbf{k}_k\right) - \mathbf{k}_i\right], \tag{9}$$

where the $\mathbf{k}_i$ are the individual momenta of the three particles. Thus, in coordinate space the wave function is written in the form:

$$\Psi = \Psi_1(\mathbf{x}_1, \mathbf{y}_1) + \Psi_2(\mathbf{x}_2, \mathbf{y}_2) + \Psi_3(\mathbf{x}_3, \mathbf{y}_3). \tag{10}$$

In momentum space one simply replaces the position variables $(\mathbf{x}_i, \mathbf{y}_i)$ by the conjugate-momentum variables $(\mathbf{p}_i, \mathbf{q}_i)$. In addition, for the momentum-space calculations it is customary to introduce the two-body t-matrices via

$$G_i(E)V_i = G_0(E)t_i, \tag{11}$$

and to write the bound-state Faddeev equations in the form:

$$|\Psi_i\rangle = G_0(E)t_i[|\Psi_j\rangle + |\Psi_k\rangle]. \tag{12}$$

For this case the three-body potential can then be included by summing the perturbation series.[11]

In both representations each Faddeev amplitude is projected onto a different complete set of channel wave functions which represent the orbital angular momentum and spin-isospin variables. These channel basis states are normally defined for the Faddeev amplitude $|\Psi_i\rangle$ as

$$|\alpha\rangle = |[(\ell_\alpha, s_\alpha)j_\alpha, (L_\alpha, S_\alpha)J_\alpha)]JM; (t_\alpha, T_\alpha)TM_T\rangle, \tag{13}$$

where

ℓ_α = relative orbital angular momentum of particles j and k;
s_α = total spin angular momentum of particles j and k;
j_α = total angular momentum of particles j and k;
L_α = orbital angular momentum of particle i relative to particles j and k;
S_α = spin of particle i $(S_\alpha = \frac{1}{2})$;
J_α = total angular momentum of particle i;
J = total angular momentum of the system $(J = \frac{1}{2})$;
t_α = total isospin of particles j and k;
T_α = isospin of particle i $(T_\alpha = \frac{1}{2})$;
T = total isospin of the system $(T = \frac{1}{2})$.

Using these channel basis states, the Faddeev amplitude in momentum space can be written in the form

$$\Psi_i(\mathbf{p}_i, \mathbf{q}_i) = \sum_{\alpha=1}^{N_c} \frac{\psi_\alpha(p_i, q_i)}{p_i q_i} |\alpha\rangle, \tag{14}$$

and the coordinate space form is

$$\Psi_i(\mathbf{x}_i, \mathbf{y}_i) = \sum_{\alpha=1}^{N_c} \frac{\psi_\alpha(x_i, y_i)}{x_i y_i} |\alpha\rangle. \tag{15}$$

where N_c is the number of channels.

In addition, the potentials are replaced by potentials projected onto the channel functions. Thus, the two-body potentials are written in the form:

$$V_i = \sum_{\alpha=1}^{N_c} \sum_{\alpha'=1}^{N_c} |\alpha'\rangle v_{\alpha'\alpha} \langle\alpha|, \tag{16}$$

where $v_{\alpha'\alpha}$ is the projection of the interaction onto the channel states, i.e.,

$$v_{\alpha'\alpha} = \langle\alpha'|V(\mathbf{x}_i)|\alpha\rangle, \tag{17}$$

and a similar expansion is used for the W_i. The channel functions form a complete set; however, for the actual numerical calculations the set must be truncated. Early calculations used only a few channels (typically 3–5) and, consequently, the results were not very accurate. In order to obtain an accurate solution to the Faddeev equations, one must use enough channel functions to accurately describe both the Faddeev amplitudes and the potentials. In practice, the number of channels is increased until the solution has converged to the desired accuracy, where the convergence is determined by the value of the bound-state energy. The difficulty of the numerical calculations increases as the number of channels is increased. For the Schrödinger equation the binding energy increases monotonically as the number of channel states is increased, but this is not the case for the Faddeev equations. The Faddeev operator is not a Hermitian operator; consequently, the equation can not be derived from a variational principle. However, the Faddeev amplitudes calculated with a fixed value of N_c can be used to construct the total wave function, and this wave function can then be used to determine a variational upper bound for the binding energy[12] whose magnitude increases monotonically with N_c.

Typically, the number of channels is chosen by choosing j_α, the maximum value of the angular momentum of the interacting pair (j, k). The initial calculations were performed using the channels with $j_\alpha \leq 1$ and positive parity; this case has five channels and corresponds to using a two-body potential which has only $^1S_0$ and $^3S_1$–$^3D_1$ interactions. Using a modern computer, calculations with $j_\alpha \leq 2$ and both positive and negative parities can be done; this case has 18 channels and yields binding energies within 100 keV of the converged result. In order to obtain a result for the binding energy with an error less than 10 keV, it is necessary to retain all channels with $j_\alpha \leq 4$, which corresponds to 34 channels. This is the largest calculation which has been done. Estimates of the contributions from the higher channels can be calculated using perturbation theory,[13] and these calculations show that the higher channels contribute less than 10 keV additional binding energy.

Numerical Algorithms

The projection onto the N_c channel functions yields N_c coupled equations for the channel wave functions. In the momentum-space calculations the resulting equation is an integral equation which contains the unknown eigenvalue for the energy in a nonlinear form. To find the eigenvalue, Eq. (12) is rewritten in the form:

$$|\Psi_i\rangle = \lambda G_0(E) t_i [|\Psi_j\rangle + |\Psi_k\rangle], \tag{18}$$

where the parameter λ has been added to the right-hand side of (18). This equation can be solved as an eigenvalue problem for λ with a fixed value of the energy. If the value assumed for the energy is the correct value, then λ will be unity. In practice, the value of the energy is varied until a value is found for which $\lambda = 1$. The momentum-space representation of Eq. (18) is an integral equation which must be solved numerically. Since all of the vectors $(\mathbf{p}_i, \mathbf{q}_i)$ lie in a plane, the integral can be reduced to a three-dimensional integral over the variables p_i, q_i, and $\hat{\mathbf{p}}_i \cdot \hat{\mathbf{q}}_i$. Since the three nucleons are identical particles, the three Faddeev amplitudes will have the same functional form; thus, for this case only one of the three Faddeev equations has to be solved numerically. Projecting Eq. (18) onto the channel basis states gives N_c coupled equations for the $\psi_\alpha(p_i, q_i)$. One of the numerical techniques used to solve these coupled equations is to approximate the two-dimensional integrals by choosing an appropriate set of quadrature points for the variables p_i and q_i. This procedure yields a large matrix equation whose eigenvalues are the values of λ and the eigenvectors are the values of the ψ at the quadrature points. Since the desired eigenvalue of $\lambda = 1$ is normally one of the smallest eigenvalues, the matrix eigenvalue equation can be computed efficiently by an iterative technique such as the power method.[5]

In coordinate space the Faddeev equation is written as a partial differential equation. This is done by multiplying Eq. (7) by the inverse of $G_i(E)$, and the resulting equation is

$$[E - (H_0 + V_i + W_i)]|\Psi_i\rangle = (V_i + W_i)[|\Psi_j\rangle + |\Psi_k\rangle], \tag{19}$$

where in coordinate space the kinetic energy operator, H_0, is a partial differential operator. Once again, for identical particles it is only necessary to solve one of the three Faddeev equations. Introducing the N_c channel basis functions leads to N_c coupled partial differential equations for the channel functions, ψ_α. The energy appears as a linear term in (17), and the equations could be rewritten as an eigenvalue problem for the energy. However, for the cases with many channels the numerical calculations become prohibitive, and a procedure similar to the momentum-space calculations is used. Equation (19) is rewritten in the form:

$$[E - (H_0 + V_i)]|\Psi_i\rangle = \lambda\{W_i|\Psi_i\rangle + (V_i + W_i)[|\Psi_j\rangle + |\Psi_k\rangle]\}, \tag{20}$$

and one solves for the eigenvalue λ. Since Eq. (20) is an elliptic partial differential equation, the boundary conditions must be specified on a closed surface in order to have a unique solution. For the bound state the boundary conditions are simple since the wave function must go to zero for large separations of any two particles. To solve Eq. (20) the differential operators can be approximated by finite differences,[8] and the resulting matrix equation can be solved using the same techniques as those used for the momentum-space calculations. Another procedure is to expand the unknown channel functions in a complete set of basis functions, such as splines, and to use the method of orthogonal collocation[14] to reduce the equations to a matrix equation.[9,15] Writing the equation in the form given in Eq. (20) yields a generalized matrix eigenvalue problem, where one of the matrices is a block diagonal matrix. The block diagonal matrix is much easier to invert than a full matrix, and one can use an iterative procedure to find the eigenvalue. A particularly efficient method is to use the Lanczos algorithm.[15]

NUMERICAL RESULTS

To demonstrate the convergence of the numerical calculations, we give in Table I a brief summary of the results for the Reid Soft Core[16] (RSC), Argonne[17] V_{14} (AV14), Urbana[18] V_{14} (UV14), Super-Soft Core(C)[19] (SSC), Paris,[20] Bonn,[21] and Nijmegen[22] potential models. The results are given for various values of N_c, the number of channels for the truncated two-body potential.

TABLE I Binding energies (in MeV) for ^{3}H calculated using various two-body potentials truncated to N_c channels. The experimental value is 8.48 MeV.

$N_c =$	5	9	18	26	34
RSC	7.02	7.21	7.23	7.34	7.35
AV14	7.44	7.57	7.57	7.67	7.67
UV14	7.46	7.52	7.49	7.54	7.53
SSC	7.62	7.77	7.75	7.82	7.82
Paris	7.31	7.42	7.39		7.47
Bonn-r	8.23	8.32	8.24		8.32
Bonn-q	8.36	8.39	8.31		
Nijmegen	7.49	7.60	7.54		7.63

All of the potentials except the Bonn underbind ^{3}H by approximately 1 MeV. From this table it can be seen that most of the binding energy comes from the first five channels, the projected potential with $j_\alpha \leq 2$ and even parity. However, to obtain a result for the binding energy which is accurate to 10 keV, it is necessary to use 34 channels. To show the contribution to the binding energy from the various projected components of the two-body potential, we show in Tables II and III the matrix elements of V_j for $j_\alpha \leq 6$ calculated with a 34-channel wave function. The terms with $j_\alpha > 4$ represent the perturbation corrections to the binding energy. From these tables one can see that the higher states contribute approximately 10 keV additional binding energy.

TABLE II Potential energies (in MeV) for the RSC potential broken down according to j (the angular momentum of the interacting pair) and parity. The total kinetic energy and the total energy are given.

j	0	1	2	3	4	5	6	Σ
$\langle V_j^+ \rangle$	−13.547	−43.859	−0.188	−0.116	−0.014	−0.009	−0.002	−57.735
$\langle V_j^- \rangle$	−0.176	0.227	−0.247	0.002	−0.006	0.000	0.000	−0.200
$\langle V_j \rangle$	−13.723	−43.632	−0.435	−0.114	−0.020	−0.009	−0.002	−57.935
$\langle T \rangle$								50.577
$\langle H \rangle$								−7.358

TABLE III Potential energies (in MeV) for the AV14 potential broken down according to j (the angular momentum of the interacting pair) and parity. The total kinetic energy and the total energy are given.

j	0	1	2	3	4	5	6	Σ
$\langle V_j^+ \rangle$	−13.590	−39.323	−0.134	−0.107	−0.008	−0.007	−0.001	−53.170
$\langle V_j^- \rangle$	−0.173	0.241	−0.242	0.004	−0.007	0.000	0.000	−0.177
$\langle V_j \rangle$	−13.763	−39.082	−0.377	−0.103	−0.015	−0.007	−0.001	−53.347
$\langle T \rangle$								45.670
$\langle H \rangle$								−7.677

In addition to the two-body forces which depend only on the coordinates of two of the nucleons in the trinucleon system, there exist other forces which depend in a nontrivial manner on the coordinates of all three particles. These three-body forces are expected to be much weaker than the two-body interactions. The three-body force is not something that is made by nature, it results from the fact that in the traditional model of the trinucleon system, some important degrees of freedom are suppressed. The neglect of these degrees of freedom can lead to the underbinding of the trinucleon system. Thus, there has been considerable interest recently in including a three-body force in the model Hamiltonian. In particular, the long-range effects of these degrees of freedom should be included; the short-range effects will be suppressed by the short-range repulsion of the two-body force. Consequently, most calculations have been concerned with pion and delta degrees of freedom.

There exist two different approaches to treating these subnuclear degrees of freedom. One approach is to treat explicitly the degrees of freedom associated with nonnucleonic components. This approach has been implemented by Hajduk and Sauer,[7] who include single Δ components in the wave function. Then the three-body forces mediated by a single Δ are implicit in the formalism. This model is much more complex because of the need to keep track of the additional wave-function components, and the numerical solution of the model equations by Hajduk and Sauer is a calculational *tour de force*. This calculation finds that the introduction of the Δ-degrees of freedom leads to an additional 0.3 MeV binding energy. The second approach is to "freeze out" the nonnucleonic degrees of freedom and to include them as separate three-body forces. This is the approach adopted by the Tucson-Melbourne[23] and Brazil[24] groups. These groups have focused on a two-pion exchange force between two nucleons as shown in Figure 2, the nonrelativistic reduction of this diagram yields the two-pion exchange three-body potential. The addition of this potential to the model Hamiltonian increases the binding energy by approximately 1.5 MeV. The reason for the large difference in the additional binding energy produced by the two approaches has not been determined.

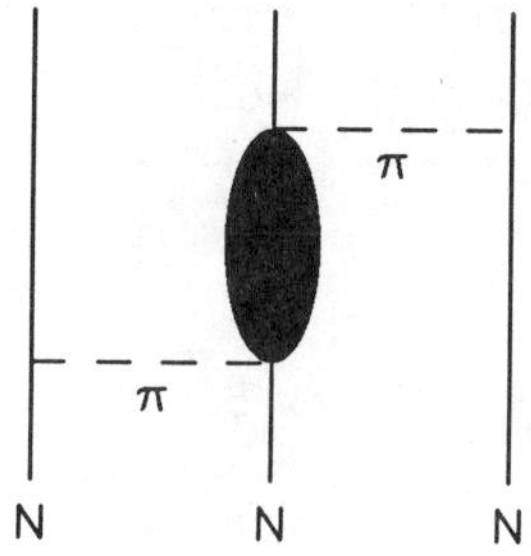

FIGURE 2. Two-pion exchange three-nucleon force.

To illustrate the effect of a three-body force, we give in Tables IV and V the results for the Tucson-Melbourne and the Brazil three-body forces used with the Reid Soft-Core two-body potential. In addition, the variational bounds calculated using the wave function constructed from the Faddeev amplitudes are given. As noted above, the binding energy does not increase monotonically as the number of channels is increased; however, the variational bound does increase. These results show that 34 channels are required to obtain an accurate value for the binding energy. Also, the addition of the three-body potential has increased the binding energy by approximately 1.5 MeV.

Once we have obtained an accurate solution to the Faddeev equations we can use the resulting wave functions to calculate additional properties of the bound-trinucleon system. To illustrate the types of calculations done, we give several examples associated with the Coulomb interaction.

TABLE IV ^{3}H binding energies (MeV) for the RSC plus the Tucson-Melbourne three-body force model and the variational bounds.

N_c	5	9	18	34
$-E_F$	7.54	8.33	8.93	8.86
$-\langle H \rangle$	8.75	8.00	8.60	8.82

TABLE V ^{3}H binding energies (MeV) for the RSC plus the Brazil three-body force model and the variational bounds.

N_c	5	9	18	34
$-E_F$	7.66	8.77	8.71	8.89
$-\langle H \rangle$	8.28	8.60	8.72	8.91

Coulomb Effects

Most investigations of the trinucleon bound state have assumed that the interactions have charge symmetry and charge independence. There are many mechanisms which can break this isospin symmetry. Primary among these is the Coulomb interaction between the two protons in ^{3}He. The mass difference between the mirror pairs ^{3}He and ^{3}H is a manifestation of this charge symmetry breaking. After subtracting the mass difference between a free proton and a free neutron, a binding energy difference of 764 keV remains. It has been known for fifty years[25] that the bulk of this difference is due to the Coulomb interaction in ^{3}He; however, only recently has a qualitative understanding been achieved.[26–28] The effects of the Coulomb interaction can be studied by solving the Faddeev equations for the ^{3}He system both with and without the Coulomb potential between the protons. We have done this using several different realistic two-body and three-body nucleon-nucleon potentials.[29] Using the results of these calculations, we have generated a data set with many different binding energies. These data were then used to study the variation of the bound-state properties with the binding energy.

The charge form factors for the trinucleon systems, $F_{ch}(q^2)$, and the associated quantities derived from them, such as the charge density

$$\rho(r) = \frac{2}{\pi} \int_0^\infty \frac{\sin qr}{qr} F_{ch}(q^2)\, q^2\, dq \tag{21}$$

and the corresponding root-mean-square charge radius $\langle r^2 \rangle^{1/2}$, where

$$\langle r^2 \rangle = \int_0^\infty r^2\, \rho(r)\, r^2\, dr, \tag{22}$$

serve as tests for the quality of the wave function derived from the model Hamiltonian. The charge form factors are measured by electron elastic scattering at momentum transfer q, and the high q values correspond to the short-range structure of the wave function. Thus, the magnitude of the secondary maxima and the position of the diffraction minima reflect the short-range correlations in the wave function. These features are also reflected in the behavior of $\rho(r)$ for small values of r. The charge radii are less sensitive functions of the short-range correlations, and we discuss these first.

In Figure 3 we have plotted the root-mean-square charge radii for both ^{3}He and ^{3}H as a function of the binding energy for various potential models. While none of the model Hamiltonians reproduce the correct binding energy, the plot shows how the charge radii scale with the binding energy for realistic interactions. From this plot one can see that the charge radii exhibit a simple functional behavior, with only a small spread of values for a fixed value of the binding energy. The Coulomb energy shift is evident in the dashed curve, and the Coulomb barrier separates the two curves by approximately 0.04 fm. The Saclay results[30] are also shown on this figure, and it can be seen that the Coulomb interaction is necessary to obtain agreement with the experimental results. If one assumes that the asymptotic part of the wave function dominates the matrix element of r^2, then the rms radius should be proportional to the square root of the binding energy. This is the case for the curves shown in Figure 3. Thus, one would expect that since the Coulomb interaction decreases the binding energy, the charge radius should increase when a Coulomb potential is present. However, there is an additional effect due to the Coulomb barrier which acts to reduce the radius.

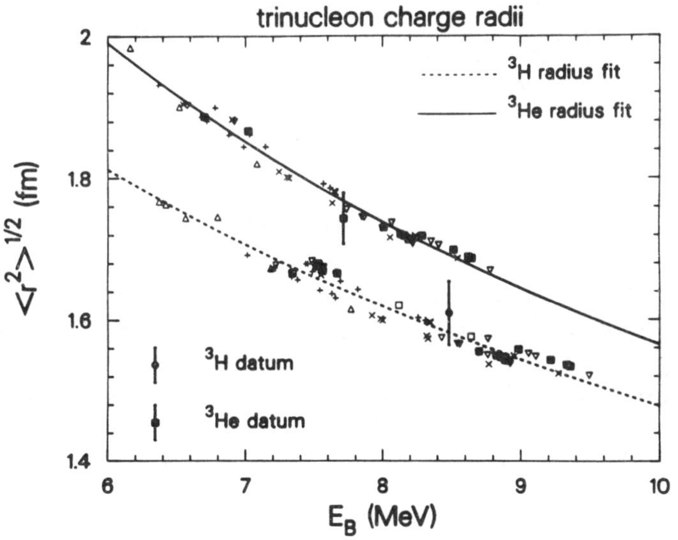

FIGURE 3. Trinucleon root-mean-square radii versus three-body binding energy for diverse potentials.

Numerical calculations for ^{3}He with and without the Coulomb interactions can be used to test the accuracy of using first-order perturbation theory to calculate the Coulomb energy. This test shows that the perturbation approximation is valid, and we have plotted in Figure 4 the Coulomb energies calculated with our model wave functions using a dipole proton form factor. The value of the Coulomb energy at the physical binding energy can be found from the fit to these model calculations, and the value is 652 keV. The estimated size of the second-order correction is approximately 4 keV. Thus, the Coulomb interaction produces most of the binding energy difference. The remaining 100 keV must be due to other effects such as meson-exchange currents or relativistic effects.

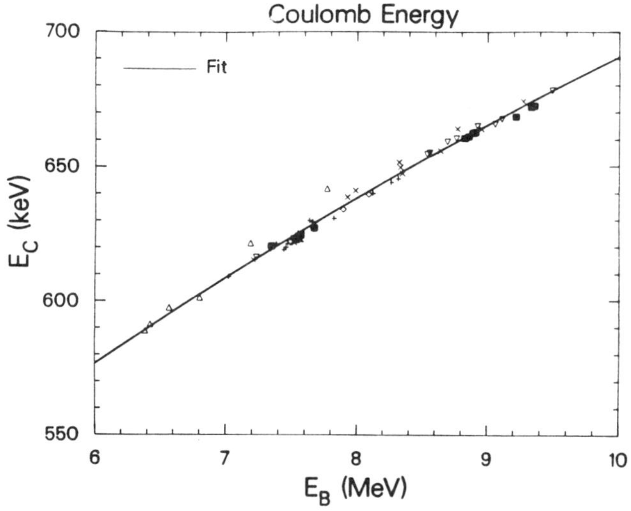

FIGURE 4. ^{3}He Coulomb energy calculated using first-order perturbation theory versus the binding energy for various potential models.

A recurring problem for ^{3}He has been the charge form factor. The experimental behavior of the $F_{ch}(q^2)$ for large q values is significantly different than the behavior predicted by the potential models for the trinucleon system. In configuration space this is reflected in the hole in the ^{3}He charge density which is not present in the model calculations. In Figure 5 the point-nucleon charge density calculated with the Reid-soft-core potential is compared with the experimental values. The experimental values shown with their error bars were generated by fitting the experimental data for the

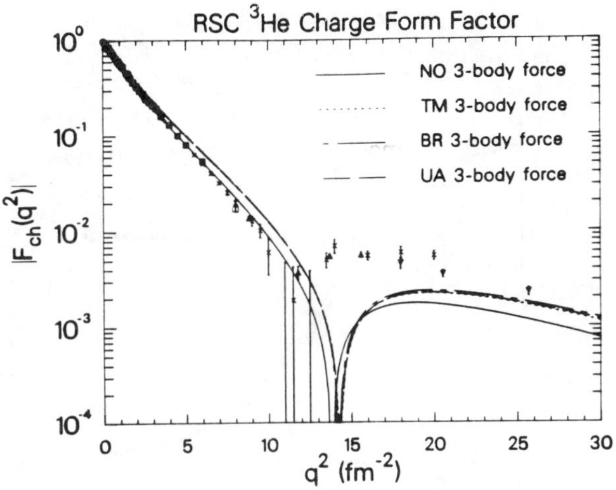

FIGURE 5. The magnitude of the RSC ^{3}He charge form factor in the impulse approximation for several three-body force models plotted versus q^2, together with the experimental data.

^{3}He form factors using a Fourier-Bessel expansion in the manner described by Friar and Negele.[31] In view of the strong-range repulsion of the two-body potential, many theorists found the lack of a hole in the theoretical density surprising. However, it has been shown that the large value of the calculated charge density at the origin is caused by the large value of the three-body wave function when the three nucleons have a linear configuration.[32] One possibility suggested for correcting this discrepancy was the addition of a three-body potential to the model Hamiltonian. The angular dependence of this interaction could reduce the probability for the linear configuration, and hence reduce the density near the origin. However, calculations with several different three-body interactions produced no significant change in the charge density. The results for several different three-body potentials are shown in Figure 6. From these plots one can see that the addition of a three-body force only produces a slight depression in the charge density for small values of r. Other effects such as meson-exchange currents are probably needed to reproduce the experimental data.

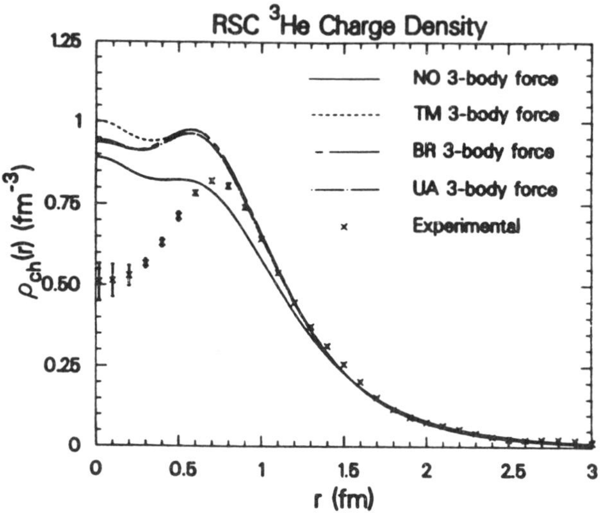

FIGURE 6. The point-nucleon charge density for ^{3}He calculated using the RSC potential and different three-body potentials. The experimental values are also shown.

SCATTERING CALCULATIONS

For the scattering problem the Faddeev equations are considerably more difficult to solve. The source of this difficulty is the fact that the Faddeev amplitudes for the scattering state are not square integrable. In momentum space this results in singularities in the kernel of the integral equation, and the numerical integration of the equation becomes more complicated. In coordinate space the asymptotic boundary conditions are more difficult to formulate in a form convenient for numerical calculations.

The Faddeev equations for the scattering problem are similar to the bound-state case, except for the addition of the incident wave. The Faddeev amplitudes are still defined as in Eq. (4), and the equations have the form

$$|\Psi_i\rangle = |\Phi_i\rangle + G_0(E)t_i[|\Psi_j\rangle + |\Psi_k\rangle], \tag{23}$$

where $|\Phi_i\rangle$ represents a state with particle i incident with relative momentum $\mathbf{q}_0$ on the bound pair (j,k). In the momentum-space calculations the transition amplitudes

$$T_{ij} = \langle\Phi_i|V^i|\Psi_j\rangle, \tag{24}$$

where $V^i = V_j + V_k$ are the quantities which are calculated instead of the wave function. To evaluate these amplitudes, one defines the transition operator U_{ij} by the relation:

$$\langle \Phi_i | U_{ij} | \Phi_i \rangle = \langle \Phi_i | V^i | \Psi_j \rangle. \tag{25}$$

This transition can be found by solving the AGS equations[5,33]

$$U_{ij} = (1 - \delta_{ij}) G_0^{-1} + \sum_{k=1}^{3} (1 - \delta_{ki}) V_k G_k U_{kj}. \tag{26}$$

Using the channel basis states one obtains a set of coupled integral equations which must be solved numerically. The integrals in these equations are two-dimensional, and the numerical solution of the equations for a realistic interaction is a large numerical calculation, even on a modern computer. Nevertheless, computer programs to solve these equations have been developed,[34–36] and these codes are being used to test the current model Hamiltonians. Since the numerical algorithms used are the same as those used for the bound-state calculations, the equations are reduced to large matrix equations which must be inverted numerically. This is normally done by using an iterative procedure known as the multiple-scattering series,[5,37] and once again the numerical calculation is reduced to matrix multiplications which are done efficiently on a vector computer. In addition, various approximation methods have been used to solve the equations. One procedure is to approximate the two-body t-matrix, t_i, by a separable expansion,[38,39] then the two-dimensional integrals are reduced to one-dimensional integrals. The one-dimensional integral equations are considerably easier to solve numerically. However, the separable expansion may require many terms to give an accurate approximation to t_i.

In coordinate space the partial-differential equations for the Faddeev amplitudes are the same as for the bound-state equations. The numerical procedures for solving the equations are the same as those used in the bound-state calculations. However, as mentioned above, the boundary conditions are different, and finding an exact form for the boundary conditions which is convenient for numerical calculations has proved to be very difficult. The initial calculations[8] used an approximate form derived by rewriting Eq. (23) in the form

$$|\Psi_i\rangle = |\Phi_i\rangle + G_i(E) V_i [|\Psi_j\rangle + |\Psi_k\rangle], \tag{27}$$

and using the stationary-phase method for the configuration-space representation of the integral equation. Using these boundary conditions, the problem is reduced to a matrix equation which can be solved by the standard methods. The initial calculations by the Grenoble group[40] seemed to indicate that this procedure could give accurate results. However, the errors introduced by the use of stationary-phase boundary conditions at a finite matching radius were difficult to determine. Recently, numerical studies[41] have indicated that the stationary approximation is only valid at distances which are prohibitive for numerical calculations. Before any serious calculations can be attempted for the Faddeev equations in coordinate space, these questions about the boundary conditions must be resolved.

SUMMARY

We have now reached the point where the Faddeev equations can be solved "exactly" for the bound-state three-body problem, either in momentum space or in coordinate space. For both the momentum-space and the configuration-space calculations, the problem can be reduced to a large matrix calculation which can be performed efficiently on present supercomputers. These solutions have become an important tool which can be used to test our understanding of the nuclear system. Using these "exact" wave functions one can reproduce many of the experimental properties of the trinucleon bound-state system. However, there exist data which depend upon the short-range correlations that are not reproduced by the traditional models. Including the effects of meson-exchange currents can remove many of the discrepancies, but the question of whether one must include other subnuclear degrees of freedom such as quarks is still not answered. Only through more detailed calculations can we determine the need for including these degrees of freedom in our model calculations. In addition, it is now known that many of the bound-state observables scale with the bound-state energy,[42] and if the model Hamiltonian reproduces the correct binding energy, the other bound-state properties are also correct. However, this is not the case for the trinucleon scattering problem. For the scattering problem the results are much more sensitive to individual components of the nuclear interactions. An example of this is the polarization data for neutron-deuteron scattering which is sensitive to the p-wave components of the interactions. Only by detailed studies of the trinucleon scattering problem can we determine the effects of the subnuclear degrees of freedom. The numerical calculations for the scattering problem are much larger than the calculations for the bound state, and the calculations are at the limit of present computers. As larger and faster computers become available, more accurate calculations can be performed. During the past ten years significant progress has been made on the trinucleon bound-state problem, and similar progress should be made on the scattering problem during the next ten years.

Acknowledgments

This work was supported, in part, by the U. S. Department of Energy. The author is grateful for the help of his collaborators J. L. Friar, B. F. Gibson, and W. N. Polyzou.

REFERENCES

1. L. D. Faddeev, Mathematical Aspects of the Three-Body Problem in Quantum Scattering Theory (Davey, New York, 1965).
2. J. Carlson, V. R. Pandharipande, and R. B. Wiringa, Nucl. Phys., A401, 59 (1983).
3. C. Ciofi delgi Atti and S. Smith, Phys. Rev. C, 32, 1090, (1984).
4. J. L. Ballot and M. Fabre de la Ripelle, Ann. Phys. (NY), 127, 62 (1980).
5. W. Glöckle, The Quantum Mechanical Few-Body Problem (Springer-Verlag, New York, 1983), and references therein.
6. E. P. Harper, Y. E. Kim, and A. Tubis, Phys. Rev. Lett., 28, 1533 (1972); W. Glöckle, G. Hasberg, A. R. Neghabian, Z. Phys., A300, 217 (1982).
7. C. Hajduk and P. U. Sauer, Nucl. Phys., A322, 329 (1979).
8. S. P. Merkuriev, C. Gignoux, and A. Laverne, Ann. Phys. (NY), 99, 30 (1976).
9. G. L. Payne, J. L. Friar, B. F. Gibson, and I. R. Afnan, Phys. Rev. C, 22, 823 (1980).
10. S. Ishikawa, T. Sasakawa, T. Sawada, and T. Ueda, Phys. Rev. Lett., 53, 1877 (1984).
11. A. Bömelburg, Phys. Rev. C, 34, 14 (1986).
12. C. R. Chen, G. L. Payne, J. L. Friar, and B. F. Gibson, Phys. Rev. C, 33, 1740 (1986).
13. J. L. Friar, B. F. Gibson, and G. L. Payne, Phys. Rev. C, 36, 1138 (1987).
14. P. M. Prenter, Splines and Variational Methods (Wiley, New York, 1975).
15. G. L. Payne, Nucl. Phys., A463, 139c (1987).
16. R. V. Reid, Jr., Ann. Phys. (NY), 50, 411 (1968).
17. R. B. Wiringa, R. A. Smith, and T. L. Ainsworth, Phys. Rev. C, 29, 1207 (1984).
18. R. Schiavilla, V. R. Pandharipande, and R. B. Wiringa Nucl. Phys., A449, 219 (1986).
19. R. de Tourreil and D.W.L. Sprung, Nucl. Phys., A201, 193 (1973).

20. M. LaCombe, B. Loiseau, J. M. Richard, R. Vinh Mauh, J. Cote, P. Pires, and R. de Tourreil, Phys. Rev. C, 21, 861 (1980).
21. R. Machleidt, K. Holinde, and C. Elster, Phys. Rep., 149, 1 (1987).
22. M. M. Nagels, T. A. Rijken, and J. J. DeSwart, Phys. Rev. D, 17, 768 (1978).
23. S. A. Coon, M. D. Scadron, P. C. McNamee, B. R. Barrett D.W.E. Blatt, and B.B.J. McKellar, Nucl. Phys., A317, 242 (1979).
24. H. T. Coelho, T. K. Das, and M. R. Robilotta, Phys. Rev. C, 28, 1812 (1983).
25. H. Bethe and R. F. Bacher, Rev. Mod. Phys., 8, 82 (1936).
26. M. Fabre de la Ripelle, Fizika, 4, 1 (1972).
27. R. A. Brandenburg, S. A. Coon, and P. U. Sauer, Nucl. Phys., A294, 305 (1978).
28. J. L. Friar and B. F. Gibson, Phys. Rev. C, 18, 908 (1978).
29. J. L. Friar, B. F. Gibson, and G. L. Payne, Phys. Rev. C, 35, 1502 (1987).
30. F.-P. Juster et al., Phys. Rev. Lett., 55, 2261 (1985).
31. J. L. Friar and J. W. Negele, Adv. Nucl. Phys., 8, 219 (1975).
32. J. L. Friar, B. F. Gibson, E. L. Tomusiak, and G. L. Payne, Phys. Rev. C, 24, 665 (1981).
33. E. O. Alt, P. Grassberger, and W. Sandhas, Nucl. Phys., B2, 167 (1967).
34. T. Takemiya, Prog. Theor. Phys., 74, 301 (1985).
35. R. A. Brandenburg, Few-Body Systems, 3, 59 (1987).
36. H. Witala, T. Cornelius, and W. Glöckle, Few-Body Systems, 3, 123 (1987).
37. R. A. Malfliet and J. A. Tjon, Nucl. Phys., A127, 161 (1969).
38. W. Plessas, J. Haidenbauer, and Y. Koike, in Theoretical and Experimental Investigations of Hadronic Few-Body Systems, edited by C. Ciofi degli Atti, O. Benhar, E. Pace, and G. Salmé (Springer-Verlag, New York, 1986).
39. Y. Koike, J. Haidenbauer, and W. Plessas, Phys. Rev. C, 35, 396 (1987).
40. J. J. Benayoun, J. Chauvin, C. Gignoux, and A. Laverne, Phys. Rev. Lett., 36, 1438 (1976); A. Laverne and C. Gignoux, Nucl. Phys., A203, 597 (1973).
41. Z. C. Kuruoglu and F. S. Levin, Phys. Rev. C, 36, 49 (1987).
42. J. L. Friar, B. F. Gibson, C. R. Chen and G. L. Payne, Phys. Rev. Lett., 161B, 241 (1985).

AUTHOR INDEX

Areti, H., 229
Atac, R., 229

Barnes, T., 83
Biel, J., 229
Blum, V., 157
Bottcher, C., 157

Chu, Shih-I, 283
Cook, A., 229
Cusson, R.Y., 157

Deppe, J., 229
deReus, T., 1

Edel, M., 229

Fink, J., 157
Fischler, M., 229
Froese Fischer, C., 117

Gaines, I., 229
Graf, O., 1
Graf, S., 1
Greiner, W., 1, 157

Hance, R., 229
Husby, D., 229

Isely, M., 229

Kalos, M.H., 57
Kirby, K.P., 269
Kogut, J., 43

Maruhn, J., 157, 173
Miranda, E., 229
Miranda, M., 229
Müller, B., 1
Müller, U., 1

Nash, T., 229
Negele, J.W., 133

Olson, R.E., 107

Payne, G.L., 305
Pham, T., 229
Pieper, S., 69

Reinhard, P.G., 157
Reinhardt, J., 1
Rufa, M., 157
Runge, K.J., 57

Sapirstein, J., 25
Schaefer, H.F., 247
Shakin, C., 207
Stevenson, D.E., 221
Strayer, M.R., 157

Vitiello, S.A., 57

Umar, S., 157

Wilets, L., 193
Wu, J., 157

Zmuda, T., 229